PREDICTION AND IMPROVED ESTIMATION IN LINEAR MODELS

"Time present and time past
Are both perhaps present in time future,
And time future contained in time past."
T. S. Eliot: Burnt Norton

JOHN BIBBY
University of St. Andrews, Scotland
and The Open University, Milton Keynes, England

and

HELGE TOUTENBURG
Central Institute for Mathematics and Mechanics of the Academy of Sciences of the German Democratic Republic

A Wiley — Interscience Publication

JOHN WILEY & SONS
Chichester · New York · Brisbane · Toronto

Library of Congress Cataloging in Publication Data:
Bibby, John.
Prediction and improved estimation in linear models.
"A Wiley Interscience publication."
Rev. and updated version of Vorhersage in linearen
Modellen by H. Toutenburg.
Bibliography: p.
Includes index.
1. Regression analysis. 2. Prediction theory.
I. Toutenburg, Helge, joint author. II. Toutenburg,
Helge. Vorhersage in linearen Modellen.
III. Title.

QA 278.2.B5 519.5′36 76-7533
ISBN 0 471 01656 X

Printed in GDR

Preface

The main aim of this book is to bring to the attention of the English-speaking public some recent theoretical work from the German Democratic Republic. It is perhaps a sad commentary on the insularity of Britain and her former colonies that this can only be done through the medium of a direct English translation. Yet it remains a fact that works appearing in German or other foreign languages often stay closed as far as English readers are concerned unless and until a translation is published.

This is particularly so with works from the German Democratic Republic (GDR). As one of the socialist countries, authors and readers in the GDR tend to look eastwards, and when we from the West look in their direction we tend to see over their heads to Moscow or further afield. It is easy to forget that in the realm of statistics alone, several top quality journals such as Biometrische Zeitschrift and Mathematische Operationsforschung und Statistik are published in the GDR. Publishers such as Akademie-Verlag also have a large statistical output of high quality — the Wissenschaftliche Taschenbücher of Akademie-Verlag are particularly noteworthy as an example of a series of economical yet up-to-date student texts which could well be copied in the West.

To return to the present text however, as was stated above its main aim is to present in a readily accessible form certain theoretical results in statistical estimation and prediction which have been developed in recent years, especially at the Central Institute for Mathematics and Mechanics of the Academy of Sciences of the GDR. The original works are referred to in the Bibliography, but much of the present book is based upon Toutenburg (1975), in which was collected many of the author's own results. The overall relationship between that text and the present one is described in Chapter 1, but some details of the origins of the individual chapters of this book may perhaps be given here.

Chapters 1—3 are newly written, and Bibby holds the main responsibility for these. Chapters 4—7 give a liberal reinterpretation of the corresponding chapters in Toutenburg's book, and were worked on by both authors jointly. TOUTENBURG holds the major responsibility for chapters 8—9, which follow quite closely the last two chapters of his book. Chapter 10 is a new addition, based on recent work by TOUTENBURG and the Russians KUKS and OLMAN. Chapter 11 was written by BIBBY. Appendix A is a translation by TOUTENBURG of his original appendix, while Appendix B is a new addition, contributed by BIBBY.

Having said the above, both authors are pleased to share responsibility for the whole book. The order of their names on the title page signifies no seniority, but merely the whims of the alphabet.

In preparing this work the authors have incurred many debts, most of them fortunately being not financial. In the GDR we are grateful to Professors HELGA and OLAF BUNKE, and to the Central Institute for Mathematics and Mechanics of the Akademie der Wissenschaften der DDR for a grant which enabled the authors to meet to finalize the manuscript of the book.

In Britain we should like to thank Ms. LESLEY BUTLER, Mr. JAMES CAMERON of JOHN WILEY & SONS, and the Travel Fund Committee of the Open University for financial assistance towards a much-needed trip to Berlin.

In conclusion, we must thank our two wives Zorina and Sabine (one each) for the passive resistance which addicted authors soon come to expect. They cannot be held responsible.

Berlin, August 1977

JOHN BIBBY HELGE TOUTENBURG

Foreword to the German Edition

(TOUTENBURG 1975*)

Linear models hold a central place in mathematical statistics as well as in various areas of statistical application such as economics, biology, and technology. With the aid of such models random processes can be described and inferences made in a realistic yet flexible and mathematically tractable manner.

The theory of linear models has been developed further than in other applied areas, and in recent years research on regression models particularly has produced a series of new ideas. Among these we may include the use of prior restrictions and auxiliary information on the parameters of the model, the extension of point estimators to estimation regions of various types, and the unified treatment of various complications and extensions which developed from the diverse requirements of different fields of application.

The problems of estimation and prediction using auxiliary information have been investigated for several years by the members of the Berlin Seminar for Mathematical Statistics under the leadership of Herrn Prof. Dr. O. BUNKE, and numerous new results have been obtained.

Because of the practical importance of linear models, and the fact that until now the only unified presentation of prediction and estimation methods was in articles of various languages, the author was glad to have the opportunity of surveying the existing literature and incorporating some of his own work in the present book. The main aim of this book is to provide a basis for the application of the techniques presented here in economics, technology, biology, agriculture, and medicine. For this reason considerable weight has been placed upon the practicability of the methods presented.

The reader should have some understanding of statistics and linear algebra, particulary matrix theory. With certain slight exceptions the

* Full reference to citations such as TOUTENBURG (1975) are given in the bibliography.

necessary theorems are proved in detail, so that the book is suitable for applied scientists, as well as practitioners and advanced students. The appendix outlines the necessary theory of linear algebra, matrix theory, and the distribution of functions of normal variables.

The author is indebted to Frau Prof. Dr. Helga Bunke, Frau Prof. em. Dr. Dr. h. c. Erna Weber, Prof. Dr. H. Ahrens, Prof. Dr. H. Bandemer, Prof. Dr. O. Bunke, Prof. Dr. M. Peschel and Doz. Dr. habil. D. Rasch for their critical comments and suggestions, many of which have resulted in changes in the text, and also to Frau M. Zander for the care which she took in preparing the manuscript.

Finally, I should like to express my thanks to Akademie-Verlag, and especially to Frau Dipl.-Math. R. Helle for her kind cooperation.

Berlin, November 1974 H. Toutenburg

Contents

1

Basic ideas

1.1 Introduction

1.1.1 Prediction, estimation and bias. Man has always aspired to see into the future. At one extreme the astrologers have tried to foretell the coming of good luck or bad, and at a more mundane level each one of us must crystal-ball gaze everytime we decide to leave the house without an umbrella.

As KENDALL (1973, p. 115) points out, the English language is rich in words which describe such attempts at futuroscopy — forecast, foretell, foresee (even foretaste), prediction, prevision, prognostication. Most commonly used in the technical literature are the words prediction and forecast. Some authors (KENDALL 1973, p. 115; BROWN 1962, p. 2) reserve the former for subjective estimates and verbal descriptions, while the latter is used for 'objective computations' of a quantifiable kind. However in this book we shall not trouble over such nice distinctions, and for our purposes prediction and forecasting will be regarded as synonymous. (See also Section 1.4.1).

The ubiquity of forecasts is obvious, as is the variability of the methods available. It is equally clear that no method of prediction is perfect, but that some may be better than others. In this book we examine and compare various methods of prediction and of estimation, placing a more critical emphasis than most authors upon the arbitrariness of the conventional criteria of 'best-ness'.

For instance, the requirement of unbiasedness is generally regarded as a *sine qua non*. While this undoubtedly leads to simpler mathematics (perhaps no trivial advantage), it is hard to justify on other grounds. And even the mathematical rationale for unbiasedness is now in doubt, as shown by STEIN (1956) and subsequent authors. More recently, methods such as ridge regression (HOERL and KENNARD 1970a, 1970b) have spread the gospel of biased estimation in a more practical context. One of the main aims of this book is to examine the conditions under which biased estimators and predictors can lead to an improvement over the con-

This may also be written $\hat{\boldsymbol{\beta}} = \boldsymbol{T}^{-1}\boldsymbol{X}'\boldsymbol{y}$ or $\hat{\boldsymbol{\beta}} = \boldsymbol{G}\boldsymbol{X}'\boldsymbol{y}$ where $\boldsymbol{T} = \boldsymbol{X}'\boldsymbol{X}$ and $\boldsymbol{G} = \boldsymbol{T}^{-1}$. When $\boldsymbol{T}$ has no regular inverse, a definition using generalized inverses may be employed.

1.2.2 Linear estimators. If $\boldsymbol{y}$ is a random vector, and $\boldsymbol{A}$ and $\boldsymbol{c}$ are a matrix and a vector of the relevant size, then the vector

$$\boldsymbol{b}_0 = \boldsymbol{A}\boldsymbol{y} + \boldsymbol{c} \tag{1.2.3}$$

may be called a 'linear estimator' of the vector $\boldsymbol{\beta}$. In the special case where $\boldsymbol{c} = \boldsymbol{0}$ we say that $\boldsymbol{b}_0$ is a 'homogeneous' estimator. When $\boldsymbol{c}$ may be non-zero, we call $\boldsymbol{b}_0$ a 'heterogeneous estimator'.

Clearly the OLSE given by (1.2.2) is a special case of a homogeneous linear estimator, obtained by putting $\boldsymbol{A} = (\boldsymbol{X}'\boldsymbol{X})^{-1}\boldsymbol{X}'$ and $\boldsymbol{c} = \boldsymbol{0}$ in (1.2.3).

1.2.3 Mean square error. The bias vector and covariance matrix (dispersion) of $\hat{\boldsymbol{\beta}}$ are easily evaluated in the model $(\boldsymbol{y}, \boldsymbol{X}\boldsymbol{\beta}, \sigma^2\boldsymbol{I})$ as

$$\boldsymbol{B}_{\hat{\beta}} = E[\hat{\boldsymbol{\beta}}] - \boldsymbol{\beta} = \boldsymbol{0}$$

and

$$\boldsymbol{V}_{\hat{\beta}} = \sigma^2\,\boldsymbol{G}\,.$$

Hence the mean square error matrix is

$$MSE_{\hat{\beta}} = E[(\hat{\boldsymbol{\beta}} - \boldsymbol{\beta})\,(\hat{\boldsymbol{\beta}} - \boldsymbol{\beta})'] = \sigma^2\boldsymbol{G}\,. \tag{1.2.4}$$

In general, if $\boldsymbol{b}_0$ is a heterogeneous linear estimator given by (1.2.3) then

$$\boldsymbol{B}_{\boldsymbol{b}_0} = E[\boldsymbol{b}_0] - \boldsymbol{\beta} = \boldsymbol{A}\boldsymbol{X}\boldsymbol{\beta} + \boldsymbol{c} - \boldsymbol{\beta} = (\boldsymbol{A}\boldsymbol{X} - \boldsymbol{I})\,\boldsymbol{\beta} + \boldsymbol{c}\,, \tag{1.2.5}$$

and

$$\boldsymbol{V}_{\boldsymbol{b}_0} = \sigma^2\boldsymbol{A}\boldsymbol{A}'\,. \tag{1.2.6}$$

The mean square error matrix of $\boldsymbol{b}_0$ is

$$MSE_{\boldsymbol{b}_0} = \sigma^2\boldsymbol{A}\boldsymbol{A}' + [(\boldsymbol{A}\boldsymbol{X} - \boldsymbol{I})\,\boldsymbol{\beta} + \boldsymbol{c}]\,[(\boldsymbol{A}\boldsymbol{X} - \boldsymbol{I})\,\boldsymbol{\beta} + \boldsymbol{c}]'\,. \tag{1.2.7}$$

This last expression is the well-known formula "mean square error equals variance plus bias-squared", extended to matrices. When $\boldsymbol{A} = (\boldsymbol{X}'\boldsymbol{X})^{-1}\boldsymbol{X}'$ and $\boldsymbol{c} = \boldsymbol{0}$, the final term in (1.2.5) is zero, and the whole expression is equivalent to (1.2.4).

Note that when $\mathbf{b}_0$ is a homogeneous estimator, so that $\mathbf{c} = \mathbf{0}$, the final term in (1.2.7) is

$$(\boldsymbol{A}\boldsymbol{X} - \boldsymbol{I})\,\boldsymbol{\beta}\boldsymbol{\beta}'(\boldsymbol{A}\boldsymbol{X} - \boldsymbol{I})'\,. \tag{1.2.8}$$

The next section and the rest of the book will examine certain special cases of the above, especially (1.2.1), (1.2.3), and (1.2.7). In general we shall seek to find estimators that minimize in some sense the value of (1.2.7).

1.3 Biased estimators

1.3.1 STEIN estimation. In 1956 CHARLES STEIN presented a theorem which may be summarized as follows: if $\boldsymbol{x} \sim N_p(\boldsymbol{\mu}, \boldsymbol{I})$, where $p > 2$, then (in the emotion-charged terminology which is so loved by statisticians) the unbiased estimator $\hat{\boldsymbol{\mu}}_1 = \boldsymbol{x}$ is inadmissible. In fact JAMES and STEIN (1961) showed that in such cases the estimator

$$\hat{\boldsymbol{\mu}}_2 = \left[1 - \frac{(p-2)}{S}\right]\boldsymbol{x} \tag{1.3.1}$$

where $S = \Sigma x_i^2$, has a mean square error which is everywhere less than that of $\hat{\boldsymbol{\mu}}_1$. (See also LINDLEY and SMITH 1972, EFRON and MORRIS 1973, and FINNEY 1974, 1975).

A related estimator, proposed by LINDLEY in the discussion following STEIN (1962), is the estimator $\hat{\boldsymbol{\mu}}_3$ whose ith element is

$$\bar{x} + \left[1 - \frac{(p-3)}{S'}\right](x_i - \bar{x})\,. \tag{1.3.2}$$

Here $p\bar{x} = \Sigma x_i$ and $S' = \Sigma(x_i - \bar{x})^2$. This has the effect of shrinking the conventional estimator $\hat{\boldsymbol{\mu}}_1$ towards the overall mean, rather than towards zero as in (1.3.1), and gives an estimator which dominates the conventional one whenever $p > 3$.

The practical implications of these results could be quite serious for, as HARDING (1972) points out "if μ_1 refers to butterflies in Brazil, μ_2 to ball-bearings in Birmingham and μ_3 to Brussel sprouts in Belgium, then an admissible estimator will cause the estimates of these three quite unrelated things to be related to each other, the largest being on the whole pulled down and the smallest pushed up." (Strictly speaking HARDING's comment is incorrect, since (1.3.2) dominates $\boldsymbol{x}$ only when $p \geqq 4$, and even then it is not admissible.) However the important thing about STEIN's work from the point of view of this book is, firstly, that both (1.3.1) and (1.3.2) are biased estimators, and, secondly, that they are derived by shrinking the conventional estimator towards zero in the case of (1.3.1) and $\bar{x}$ in the case of (1.3.2).

As an example to illustrate the application of the STEIN estimator, EFRON and MORRIS (1973) give the data presented in Figure 1.3.1. The first row, $\boldsymbol{x}$, represents the batting average of fourteen baseball players after 45 bats in the 1970 season. These elements would give the maximum likelihood estimators of each player's true binomial probability of getting a hit. The second row gives a STEIN estimator, obtained by applying an arcsine transformation to each element of $\boldsymbol{x}$, using (1.3.2) on the transformed data, and then applying the inverse of the arcsine transformation. This estimator shrinks each element of $\boldsymbol{x}$ towards the overall mean

	1	2	3	4	5	6	7
$\boldsymbol{x}$	0.400	0.378	0.356	0.333	0.311	0.311	0.289
STEIN	0.303	0.299	0.294	0.289	0.284	0.284	0.278
$\boldsymbol{\mu}$	0.346	0.298	0.276	0.221	0.273	0.270	0.263

	8	9	10	11	12	13	14
$\boldsymbol{x}$	0.244	0.222	0.222	0.222	0.222	0.200	0.178
STEIN	0.268	0.263	0.263	0.263	0.263	0.257	0.251
$\boldsymbol{\mu}$	0.269	0.264	0.256	0.304	0.264	0.285	0.319

Figure 1.3.1 Application of the JAMES-STEIN estimator to predicting batting averages of 14 baseball players. Source: EFRON and MORRIS (1973).

$\bar{x} = 0.278$. In the final row of the table is the 'true' vector $\boldsymbol{\mu}$ containing each player's binomial probability, as estimated at the end of the season after about 450 bats. It can be seen that the STEIN estimator is closer than x_i to μ_i in all fourteen cases, the ratio of the sum of squared errors being 0.20, even though in this case the normality assumption underlying the STEIN estimator is not satisfied. Hence we see the mean square advantage of the STEIN estimator as well as a certain robustness. (In the discussion Professor PLACKETT sceptically notes that the use of $\bar{x}$ as an estimate of each μ_i would lead to an even greater improvement, and is somewhat unfairly put down by the riposte that "one estimate doth not an estimator make". This baseball example was also followed up by STONE (1974, p. 120), who proposed an alternative prediction—based rationale for biased techniques.) For BAYESIAN and minimal mean square error interpretations of the STEIN estimator, see ZELLNER and VANDAELE (1975, p. 633).

1.3.2 Ridge regression. Although more recent than the STEIN estimators discussed in the previous section, the family of ridge estimators proposed by HOERL and KENNARD (1970 a, b) is perhaps the first biased procedure to have found widespread practical application. This may not be completely unconnected with the fact that its theoretical basis is somewhat unsatisfactory.

As an extension of (1.2.2), HOERL and KENNARD (1970a) proposed the so-called 'ridge estimator', defined by

$$\boldsymbol{\beta}_k = (\boldsymbol{X}'\boldsymbol{X} + k\boldsymbol{I})^{-1}\,\boldsymbol{X}'\boldsymbol{y}\,. \tag{1.3.3}$$

Clearly (1.2.2) is the special case of (1.3.3) obtained by putting $k = 0$. We may also write (1.3.3) as

$$\boldsymbol{\beta}_k = \boldsymbol{T}_k^{-1}\boldsymbol{X}'\boldsymbol{y} = \boldsymbol{G}_k\boldsymbol{X}'\boldsymbol{y}\,,$$

where

$$T_k = X'X + kI \quad \text{and} \quad G_k = T_k^{-1} .$$

BIBBY (1972) considered generalizations of (1.3.3) which included the estimator

$$\beta(K) = (X'X + K)^{-1} X'y ,$$

where K is any matrix of the relevant size, and also

$$\beta(k, l) = [X'X + kI + l(X'X + kI)^{-1}] X'y ,$$

where k and l are two arbitrary scalars. GOLDSTEIN and SMITH (1974, p. 289) suggest the further generalization

$$[(X'X)^m + kI]^{-1} (X'X)^{m-1} X'y .$$

HOERL and KENNARD (1975) indicate that this generalization was first suggested in 1964, and that it is a special case of $\beta(K)$. To return therefore to (1.3.3), we note that

$$G_k X'X = (X'X + kI)^{-1} X'X = I - kG_k .$$

Using this we may write the expectation of the ridge estimator as

$$E[\beta_k] = G_k X'X\beta = \beta - kG_k\beta .$$

Also the variance-covariance matrix is

$$V[\beta_k] = \sigma^2 G_k X'X G_k .$$

(These expressions are special cases of (1.2.5) and (1.2.6) respectively.) The mean square error matrix of β_k is, by analogy with (1.2.7),

$$M = \sigma^2 G_k X'X G_k + k^2 G_k \beta\beta' G_k . \tag{1.3.4}$$

GOLDSTEIN and SMITH (1974) showed that nothing is lost in generality by restricting one's attention to the case where $X'X$ is diagonal. Hence we suppose that

$$X = \begin{bmatrix} \Lambda \\ 0 \end{bmatrix} , \tag{1.3.5}$$

where $\lambda_1^2, \dots, \lambda_p^2$ are the eigenvalues of $X'X$. The X matrix would be ill-conditioned when one of these eigenvalues is small, in which case the corresponding OLS estimator would have high variance, σ^2/λ_i^2.

Inserting (1.3.5) in (1.3.4) we find that

$$m_{ii} = \frac{\sigma^2\lambda_i^2 + k^2\beta_i^2}{(\lambda_i^2 + k)^2} \tag{1.3.6}$$

and

$$m_{ij} = \frac{k^2\beta_i\beta_j}{(\lambda_i^2 + k)(\lambda_j^2 + k)} , \quad \text{when } i \neq j . \tag{1.3.7}$$

Note that when $k = 0$ these expressions correspond to the OLS estimates using the canonical form. The value of the correlation coefficient,

$$r_{ij}^2 = \frac{k^4\beta_i^2\beta_j^2}{(\sigma^2\lambda_i^2 + k^2\beta_i^2)(\sigma^2\lambda_j^2 + k^2\beta_j^2)},$$

also reflects the uncorrelatedness of OLS estimates, and in addition shows that if $\lambda_i = \lambda_j = 0$, then a linear relationship exists between the ridge estimators, whatever the value of k.

However the original motivation of the ridge estimator stemmed from the observation that

$$tr\,\boldsymbol{M} = \Sigma m_{ii} = \Sigma \frac{\sigma^2\lambda_i^2 + k^2\beta_i^2}{(\lambda_i^2 + k)^2}$$

is a function which starts from $\sum \sigma^2/\lambda_i^2$ when $k = 0$, then descends to a minimum for positive k, and only later starts increasing again. In other words, for a range of positive k the OLS estimator can be 'improved' using the ridge technique. For further discussions of ridge regression see Section 10.1 and also SMITH and GOLDSTEIN (1975). For applications to prediction see BROWN (1974) and BROWN and PAYNE (1975).

1.3.3 Exclusion of meaningless estimators. It may be by the nature of a particular problem that the elements of $\boldsymbol{\beta}$ are bound to lie in some particular region — for instance, in the context of an economic production function they might have to be positive. In this case an unbiased estimator such as the OLSE may nevertheless lead to 'nonsense' estimates — negative values in the example cited above. In such situations the MSE can be reduced and the meaningfulness increased by introducing some bias and excluding the 'nonsense' values. This may be done either by substituting for them some value on the boundary of the meaningful region, or alternatively by reflecting them in some manner into the interior of this region.

SUBRAHMANYA (1970) examined such a problem in the univariate context. He had a nonnegative parameter θ, and an unbiased estimator $\tilde{\theta}$ which could however take negative values. SUBRAHMANYA then compared the properties of the following three estimators.

$$\tilde{\theta}_1 = \tilde{\theta}\,,$$

$$\tilde{\theta}_2 = \begin{cases} \tilde{\theta} & \text{if } \tilde{\theta} \geqq 0 \\ 0 & \text{if } \tilde{\theta} < 0\,, \end{cases}$$

and

$$\tilde{\theta}_3 = \begin{cases} \tilde{\theta} & \text{if } \tilde{\theta} \geqq 0 \\ -\tilde{\theta} & \text{if } \tilde{\theta} < 0\,. \end{cases}$$

Note that $\tilde{\theta}_2$ involves substituting for meaningless values of $\tilde{\theta}$ a value on the boundary of the meaningful region, while $\tilde{\theta}_3$ involves a 'reflection' into the interior of that region.

SUBRAHMANYA examined the bias, variance, and MSE of $\tilde{\theta}_1$, $\tilde{\theta}_2$ and $\tilde{\theta}_3$. His work may be simplified by noting that all three estimators take the form

$$a\tilde{\theta} + b\,|\tilde{\theta}|\,,$$

where (a, b) equals $(1, 0)$ for $\tilde{\theta}_1$, $(\frac{1}{2}, \frac{1}{2})$ for $\tilde{\theta}_2$, and $(0, 1)$ for $\tilde{\theta}_3$.

1.4 Prediction

1.4.1 Discussion. For a general survey of applied prediction problems the reader is referred to the articles on 'Prediction' and 'Prediction and forecasting, economic' in the International Encyclopedia of Social Science. These are written by KARL SCHUESSLER and VICTOR ZARNOWITZ respectively, and both include substantial bibliographies. WOLD (1963) also discusses similar points, and AITCHISON and THATCHER (1964) consider various theoretical and philosophical issues.

SCHUESSLER defined (sociological) prediction as "a stated expectation about a given aspect of social behaviour that may be verified by subsequent observation", while ZARNOWITZ says that "A forecast can be defined generally as a statement about an unknown and uncertain event — most often, but not necessarily a future event." In short, the everyday usage of the word encapsulates the essence of its technical meaning, except possibly that questions of timing are largely irrelevant. (A 'postdiction' is still a prediction in the sense to be used here.)

However the important aspect of prediction — and the way in which it differs from an estimation — is that it concerns a guess about a random event and not about a fixed parameter. Of course this introduces no new concept within the BAYESIAN paradigm. But the classical concepts of estimation such as bias and mean square error must have their definitions extended before they can be applied to prediction — see section 1.5.3.

Several authors have suggested competing taxonomies of forecasting procedures, including the late CHARLES S. ROOS in BROWN (1962, p. 5). ZARNOWITZ (1968, p. 427) suggests the distinction between *extrapolations*, on the one hand, which are based solely on the past and current values of the variables being predicted, and *causal forecasts* on the other hand, which rely also on postulated or observed relationships between the variables being predicted and other variables.

Alternatively, forecasts can be categorized by the extent to which formal methods are used. This establishes at least in principle a finely graduated spectrum ranging from informal hunches or 'guesstimates' at one extreme, to predictions based upon fully specified and strictly implemented statistical models at the other. (BROWN 1962 makes this distinction central to his exposition.)

Thirdly, one may distinguish between forecasts constructed from a single source, and those derived as weighted or unweighted averages of different predictions made by several or many individuals or organisations.

As ZARNOWITZ (1968, p. 427) points out, these various classes of forecast overlap and can be combined in various ways. "For example, a forecast of next year's GNP and its major components by a business economist may consists of any or all of the following ingredients: (1) extrapolation, of some kind, of the past behaviour of the given series; (2) relation of the series to be predicted to known or estimated values of some other variables; (3) other external information considered relevant, such as a survey of investment intentions or a government budget estimate; and (4) the judgement of the forecaster. Also, it should be noted that a group forecast, say an opinion poll, will incorporate as many different techniques as are used by the different respondents."

1.4.2 Standard forecasting methods. The standard procedures of forecasting are well surveyed by NEWBOLD and GRANGER (1974), who start from the simplest form of exponential smoothing. This attempts to estimate the trend-value of a time-series by averaging the most recent observation with the previous estimate of trend. An observed series X_t is thus converted into a "smoothed" series $\overline{X}_t$ given by

$$\overline{X}_t = \alpha X_t + (1 - \alpha)\, \overline{X}_t \qquad (0 < \alpha < 1)\,.$$

Alternatively, $\overline{X}_t$ may be written

$$\overline{X}_t = \alpha \Sigma (1 - \alpha)^i\, X_{t-i}\,,$$

the exponentially decreasing weights giving a clue to the origin of the name. The latest smoothed value is then used to give the exponential smoothing forecast of all future values.

That is,

$$\hat{X}_n(h) = \overline{X}_n \qquad h = 1, 2, 3, \ldots$$

A modification of this technique devised by HOLT (1957) and WINTERS (1960) estimates X_{n+h} by

$$\hat{X}_n(h) = \overline{X}_n + hT_n\,,$$

where T_n is an estimate of trend given by

$$T_n = c(\overline{X}_n - \overline{X}_{n-1}) + (1 - c)\, T_{n-1}\,, \qquad 0 < c < 1\,.$$

The smoothing constant c may be determined by finding the value which would have performed best on past data. A further extension of this method allows one to introduce multiplicative seasonal components also.

An alternative method of prediction is the stepwise autoregression procedure, which uses models of the form

$$X_t = \alpha + \Sigma\beta_j X_{t-j} + \text{error}\,,$$

choosing the optimal lags using arbitrary F-ratios as in stepwise regression.

Autoregression methods lead on to the whole panoply of BOX-JENKINS techniques, which will now be described.

1.4.3 The work of Box and JENKINS. Although this book is largely theoretical, it would clearly be wrong to ignore what is undoubtedly the most important recent innovation in the realm of applied forecasting, namely the book by GEORGE BOX and GWILYM JENKINS (1970). As stated in their preface, the main objective of the BOX-JENKINS technique is "to derive models possessing maximum simplicity and the minimum number of parameters consonant with representational adequacy". Note however that the use of the word 'model' here really implies little more than a particular descriptive formula — there is no pretence in the BOX-JENKINS approach that the model finally chosen is in any sense 'real', unlike the models used elsewhere in the present book, which represent the assumed structures and parameters of particular 'black boxes'. On the contrary, BOX and JENKINS have no assumed 'black box'. Their models are descriptive rather than analytic or causal, and their main aim is for parsimony rather than precision.

It may be argued that the BOX-JENKINS approach is more honest concerning what it is valid to assume about the real world. Certainly their iterative scheme, illustrated in Figure 1.4.1 is a convincing summary of

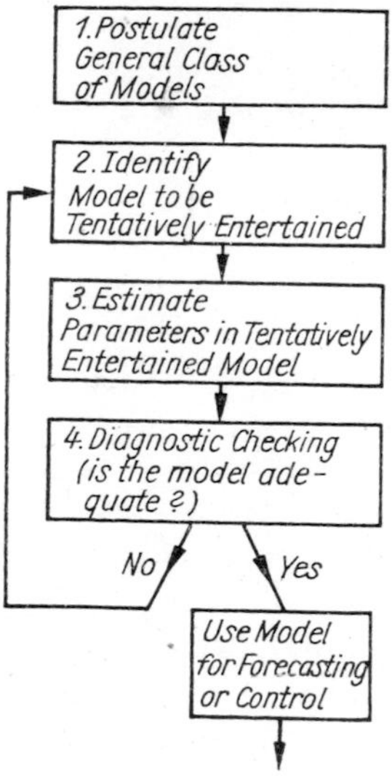

Figure 1.4.1 Stages in the iterative approach to model building. Source: BOX and JENKINS (1970).

POPPERIAN philosophy applied to the problem of forecasting and control. Box and JENKINS (1970, pp. 18—19) describe this schema as follows.

"(1) From the interaction of theory and practice, a *useful class of models* for the purposes at hand is considered.

(2) Because this class is too extensive to be conveniently fitted directly to data, rough methods for *identifying* subclasses of these models are developed. Such methods of model identification employ data and knowledge of the system to suggest an appropriate parsimonious subclass of models which may be tentatively entertained. In addition, the identification process can be used to yield rough preliminary estimates of the parameters in the model.

(3) The tentatively entertained model is *fitted* to data and its parameters *estimated*. The rough estimates obtained during the identification stage can now be used as starting values in more refined iterative methods for estimating the parameters.

(4) *Diagnostic checks* are applied with the object of uncovering possible lack of fit and diagnosing the cause. If no lack of fit is indicated, the model is ready to use. If any inadequacy is found, the iterative cycle of identification, estimation, and diagnostic checking is repeated until a suitable representation is found."

The 'useful class' of models used by Box and JENKINS in stage (1) consists of models of the form

$$\alpha(B)\,(1-B)^d\,u_t = \beta(B)\,\varepsilon_t\,,$$

where $u_1, u_2, \ldots$ are the observations made, B is the backward shift operator such that $Bu_t = u_{t-1}$, and $\alpha(.)$ and $\beta(.)$ are polynomials of order p and q. In practice, as KENDALL (1973, p. 123) points out, the parameters p, q and d are usually taken as either 1 or 2, but even so the process of estimation is rather involved.

Their first step is to difference the series of observations until stationarity is obtained in the first two moments. This leads to an estimate of d. The problem then reduces to estimating the coefficients in an autoregressive moving average (ARMA) model

$$\alpha(B)\,v_t = \beta(B)\,\varepsilon_t\,,$$

where $v_t = (1-B)^d\,u_t$ is the d-th difference of the original series. These ARMA coefficients may be obtained iteratively, first obtaining estimates of the coefficients in $\alpha(.)$, then using these to give estimates for $\beta(.)$, then obtaining revised estimates for $\alpha(.)$, and so on.

KENDALL (1973, p. 125) refers to an unpublished paper by D. J. REID who compared the BOX-JENKINS approach with other practical procedures due to BROWN (1962), WINTERS (1960), and HARRISON (1965). REID took

113 time-series, including annual, quarterly and monthly data, mostly on macro-economic variables in the United Kingdom, and examined the one-step ahead forecasting error given by each method. The BOX-JENKINS method was best in 76 of the 113 time-series. However for some sorts of series the BOX-JENKINS method appeared not to be recommended — see the tree diagram entitled 'A method of choosing a time-series prediction technique' on p. 127 of KENDALL (1973), and criticisms of this diagram on pp. 157 and 160 of the discussion on NEWBOLD and GRANGER (1974).

NEWBOLD and GRANGER (1974) also give many useful comments on the practical problems associated with forecasting. Of particular interest is their study of ways of combining different forecasting methods which lead to impracticable procedures, like those presented in this book. In conclusion, the authors propose a general set of forecasting guidelines, which may be summarised as follows:

(a) For short time series (less than 30 observations), use an exponential smoothing predictor.

(b) For series with 30—50 observations, use a stepwise autoregressive procedure, possibly combined with a HOLT-WINTERS forecast.

(c) For series containing over 50 observations, use BOX-JENKINS or a combination of this with HOLT-WINTERS and step autoregressive forecasts.

(d) Never follow guidelines (a)—(c) uncritically! In many practical situations one knows something of value about the series under consideration. This information, should, if possible, be employed in any decision. (NEWBOLD and GRANGER, 1974).

1.5 Differences between estimation and prediction

1.5.1 The OLS predictor. Suppose that we have the model $(\boldsymbol{y}, \boldsymbol{X\beta}, \sigma^2\boldsymbol{I})$, that is the special case of (1.2.1) given by

$$\boldsymbol{y} = \boldsymbol{X\beta} + \boldsymbol{u}\,, \qquad \boldsymbol{u} \sim (0, \sigma^2\boldsymbol{I})\,. \tag{1.5.1}$$

Suppose also that we have a known matrix $\boldsymbol{X}_*$, which consists of further observations on the same set of explanatory variables, and that we wish to estimate the corresponding vector of dependent variables, $\boldsymbol{y}_*$, given by

$$\boldsymbol{y}_* = \boldsymbol{X}_*\boldsymbol{\beta} + \boldsymbol{u}_*\,, \tag{1.5.2}$$

where $\boldsymbol{\beta}$ is the same vector as in (1.5.1). One obvious approach is to take the OLS estimator $\hat{\boldsymbol{\beta}}$ from (1.2.2), and use the predictor

$$\hat{\boldsymbol{y}}_* = \boldsymbol{X}_*\hat{\boldsymbol{\beta}} = \boldsymbol{X}_*\boldsymbol{\beta} + (\boldsymbol{X}'\boldsymbol{X})^{-1}\boldsymbol{X}'\boldsymbol{u}\,. \tag{1.5.3}$$

This predictor will be called the OLS predictor of $\boldsymbol{y}_*$. Its expectation is

$$E[\hat{\boldsymbol{y}}_*] = \boldsymbol{X}_* E[\hat{\boldsymbol{\beta}}] = \boldsymbol{X}_* \boldsymbol{\beta} .$$

Since this equals the expectation of $\boldsymbol{y}_*$ given by (1.5.2) we may say that $\hat{\boldsymbol{y}}_*$ is unbiased for $E[\boldsymbol{y}_*]$. Its dispersion matrix is

$$\boldsymbol{V}_{\hat{\boldsymbol{y}}_*} = \boldsymbol{X}_* \boldsymbol{V}_{\hat{\boldsymbol{\beta}}} \boldsymbol{X}'_* = \sigma^2 \boldsymbol{X}_* (\boldsymbol{X}'\boldsymbol{X})^{-1} \boldsymbol{X}'_* . \tag{1.5.4}$$

This measures the MSE of $\hat{\boldsymbol{y}}_*$ from its mean $\boldsymbol{X}_* \boldsymbol{\beta}$. However a more useful measure of the suitability of $\hat{\boldsymbol{y}}_*$ is given by its mean square error of prediction (MSEP) matrix, measured from the value that $\hat{\boldsymbol{y}}_*$ is designed to predict. This is

$$MSEP = E(\hat{\boldsymbol{y}}_* - \boldsymbol{y}_*)(\hat{\boldsymbol{y}}_* - \boldsymbol{y}_*)' . \tag{1.5.5}$$

This can be shown to equal

$$\boldsymbol{V}_{\hat{\boldsymbol{y}}_*} + \boldsymbol{V}_{\boldsymbol{y}_*} - \boldsymbol{C} - \boldsymbol{C}' , \tag{1.5.6}$$

where

$$\boldsymbol{C} = E[(\hat{\boldsymbol{y}}_* - \boldsymbol{X}_* \boldsymbol{\beta})(\boldsymbol{y}_* - \boldsymbol{X}_* \boldsymbol{\beta})'] . \tag{1.5.7}$$

If $\boldsymbol{u}_*$ in (1.5.2) is uncorrelated with $\boldsymbol{u}$ in (1.5.1), then $\boldsymbol{C}$ in (1.5.6) is a zero matrix. Hence in this case

$$\boldsymbol{M} = \boldsymbol{V}_{\hat{\boldsymbol{y}}_*} + \boldsymbol{V}_{\boldsymbol{y}_*} .$$

Now the question arises as to whether any other predictor exists which leads to a MSEP which is less than that of the OLS predictor. The work of Goldberger (1962), outlined in the next section, shows that in certain cases such predictors, which are better than OLS, do indeed exist.

1.5.2 Best linear unbiased prediction. We consider the model expressed by (1.5.1) and (1.5.2), with the additional conditions on $\boldsymbol{u}_*$ that

$$\boldsymbol{u}_* \sim (\boldsymbol{0}, \sigma^2 \boldsymbol{I}) \tag{1.5.8}$$

and

$$E(\boldsymbol{u}_* \boldsymbol{u}') = \sigma^2 \boldsymbol{W} . \tag{1.5.9}$$

(If the dispersion matrices of $\boldsymbol{u}$ or $\boldsymbol{u}_*$ are not diagonal, they can easily be brought into diagonal form by a suitable premultiplication. Note that the identity matrix in (1.5.1) is not necessarily of the same order as that in (1.5.8) — usually the latter is much smaller.)

Now consider the general linear estimator

$$\hat{\boldsymbol{y}}_0 = \boldsymbol{A}\boldsymbol{y} = \boldsymbol{A}\boldsymbol{X}\boldsymbol{\beta} + \boldsymbol{A}\boldsymbol{u} . \tag{1.5.10}$$

This has expectation

$$E[\hat{\boldsymbol{y}}_0] = \boldsymbol{A}\boldsymbol{\beta} \tag{1.5.11}$$

and dispersion matrix

$$\boldsymbol{V}_{\hat{\boldsymbol{y}}_0} = \sigma^2 \boldsymbol{A}\boldsymbol{A}' .$$

In a similar manner to (1.5.5) and (1.5.6) we may deduce if $\hat{\boldsymbol{y}}_0$ is unbiased that its MSEP is

$$MSEP_{\hat{\boldsymbol{y}}_0} = \sigma^2 \boldsymbol{A}\boldsymbol{A}' + \sigma^2 \boldsymbol{I} - \boldsymbol{C}_0 - \boldsymbol{C}_0' , \qquad (1.5.12)$$

where

$$\begin{aligned}\boldsymbol{C}_0 &= E[(\hat{\boldsymbol{y}}_0 - \boldsymbol{X}_* \boldsymbol{\beta}) (\boldsymbol{y}_* - \boldsymbol{X}_* \boldsymbol{\beta})'] \\ &= E[((\boldsymbol{A}\boldsymbol{X} - \boldsymbol{X}_*) \boldsymbol{\beta} + \boldsymbol{A}\boldsymbol{u}) \boldsymbol{u}_*'] .\end{aligned}$$

Using (1.5.8) and (1.5.9) this equals $\sigma^2 \boldsymbol{A}\boldsymbol{W}'$. Therefore substituting in (1.5.12),

$$MSEP_{\hat{\boldsymbol{y}}_0} = \sigma^2(\boldsymbol{A}\boldsymbol{A}' + \boldsymbol{I} - \boldsymbol{A}\boldsymbol{W}' - \boldsymbol{W}\boldsymbol{A}') . \qquad (1.5.13)$$

Goldberger (1962, p. 370) examined the special case of this problem where only one value of the dependent variable is to be estimated. Hence $\boldsymbol{W} = \boldsymbol{w}'$ in (1.5.9) is now a vector, and in place of (1.5.10), Goldberger took the estimate $\hat{y}_0 = \boldsymbol{c}'\boldsymbol{y}$, where $\boldsymbol{c}$ is a vector of the relevant length. As a special case of (1.5.13) this formulation leads to the mean square error of prediction

$$MSEP = \sigma^2(\boldsymbol{c}'\boldsymbol{c} + 1 - 2\boldsymbol{c}'\boldsymbol{w}) . \qquad (1.5.14)$$

Now condition (1.5.11) for $\hat{y}_0$ to be unbiased is

$$\boldsymbol{X}'\boldsymbol{c} = \boldsymbol{x}_* . \qquad (1.5.15)$$

Hence the problem of finding the 'best linear unbiased predictor' becomes one of minimizing (1.5.14) subject to (1.5.15). Goldberger shows that the optimal value of $\boldsymbol{c}$ is given by

$$\begin{aligned}\hat{\boldsymbol{c}} &= \boldsymbol{x}_*'\boldsymbol{G}\boldsymbol{X}'\boldsymbol{y} + \boldsymbol{w}'\boldsymbol{y} - \boldsymbol{w}'\boldsymbol{X}\boldsymbol{G}\boldsymbol{X}'\boldsymbol{y} \\ &= \boldsymbol{x}_*'\hat{\boldsymbol{\beta}} + \boldsymbol{w}'\boldsymbol{y} - \boldsymbol{w}'\boldsymbol{X}\hat{\boldsymbol{\beta}} \\ &= \boldsymbol{x}_*'\hat{\boldsymbol{\beta}} + \boldsymbol{w}'\hat{\boldsymbol{u}} ,\end{aligned} \qquad (1.5.16)$$

where $\hat{\boldsymbol{u}} = \boldsymbol{y} - \boldsymbol{X}\hat{\boldsymbol{\beta}}$ is the vector of OLS residuals — see (1.2.2) and (3.3.4).

To interpret (1.5.16) note that the first term is the OLS predictor corresponding to (1.5.3), and that the second term, as Goldberger puts it, "utilizes a priori knowledge of the interdependence of disturbances along with the sample residuals (which are estimates of the sample disturbances) to estimate the prediction disturbance u_*" (Goldberger 1962, p. 371).

Goldberger also examined the special case of his result when the disturbances follow a first order autoregressive process, and showed that the gain in efficiency due to using best linear unbiased prediction as compared with OLS prediction is

$$\sigma^2 \boldsymbol{w}'(\boldsymbol{I} - \boldsymbol{X}\boldsymbol{G}\boldsymbol{X}') \boldsymbol{w}$$

It is a somewhat surprising result that if several dependent variables are being predicted at the same time, then there is no way using GOLDBERGER's procedure in which one dependent variable can be used to improve the prediction of the others.

1.5.3 Estimating random variables. As suggested in Section 1.4.1, the difference between estimation and prediction comes down essentially to the question, what are we trying to estimate? Or in the notation of (1.5.2), are we concerned with predicting $\boldsymbol{y}_*$, or are we estimating its mean, $E[\boldsymbol{y}_*]$? The conventional approach would answer that we are estimating the mean, $E[\boldsymbol{y}_*]$. We however wish to estimate $\boldsymbol{y}_*$ itself. The difference may be illustrated for simplicity using two scalar random variables, p and a. For concreteness we may regard p as the prediction and a as the actual observed value. Now the MSE criterion is

$$E[(p-a)^2] = E[\{(p-\mu_p)-(a-\mu_a)+(\mu_p-\mu_a)\}^2],$$

where μ_a and μ_p are the corresponding means. Expanding the curly brackets this equals

$$Var\,[p] + Var\,[a] + (\mu_p-\mu_a)^2 - 2\,covar\,(p, a)\,. \qquad (1.5.17)$$

We may compare this with the decomposition based on deviations between p and the mean of a, which is

$$\begin{aligned} E[(p-\mu_a)^2] &= E[\{(p-\mu_p)+(\mu_p-\mu_a)\}^2] \\ &= Var\,[p] + (\mu_p-\mu_a)^2\,. \end{aligned} \qquad (1.5.18)$$

This is just the well-known "MSE equals variance plus bias-squared" formula, which omits the second and fourth terms in (1.5.17).

Expressed in a nutshell, whereas OLS prediction uses a loss function based on (1.5.18), we are concerned with (1.5.17).

1.5.4 The work of HENRI THEIL. The approach outlined above has some conceptual affinities with the post hoc method for assessing predictions advanced by THEIL (1966, ch. 2). The difference is essentially that while we are looking at ways of comparing predictors without actually examining the predictions to which they lead, THEIL's method had a spirit of 'suck it and see', and asks how a particular strategy would have fared if it had been used in the past.

Theil uses $\boldsymbol{P}$ and $\boldsymbol{A}$ to denote vectors of predicted and actual values rather than $\hat{\boldsymbol{y}}$ and $\boldsymbol{y}_*$. We shall adopt his notation here. (In fact THEIL tends to talk of predicted and actual *changes*, but nothing is lost if we think in terms of actual values.) The mean square prediction error is defined as

$$MSPE = \frac{1}{n}\sum_{i=1}^{n}(P_i - A_i)^2\,, \qquad (1.5.19)$$

where P_i is the predicted value in time period i, and A_i is the observed value. This may also be written as

$$MSPE = \frac{1}{n} \Sigma [(\bar{P} - \bar{A}) + (P_i - P) - (A_i \quad \bar{A})]^2$$

$$= (\bar{P} - \bar{A})^2 + s_P^2 + s_A^2 - 2rs_P s_A \, , \qquad (1.5.20)$$

where

$$\bar{A} = \frac{1}{n} \Sigma A_i \, , \qquad s_A^2 = \frac{1}{n} \Sigma (A_i - \bar{A})^2 \, ,$$

$$\bar{P} = \frac{1}{n} \Sigma P_i \, , \qquad s_P^2 = \frac{1}{n} \Sigma (P_i - \bar{P})^2 \, ,$$

and

$$r = \frac{\frac{1}{n} \Sigma (P_i - \bar{P}) (A_i - \bar{A})}{s_P s_A} \, .$$

This decomposition may be regarded as a sample based counterpart of (1.5.17). It has several convenient interpretations. The first term in (1.5.20) is zero when $\bar{P} = \bar{A}$ i.e. when the average predicted value coincides with the average realized value. Errors which lead to a positive value for this term may be called errors in central tendency. The final three terms in (1.5.20) represent the sample variance of $(\boldsymbol{P} - \boldsymbol{A})$. They may be written in two different ways. One leads to the decomposition

$$MSPE = (\bar{P} - \bar{A})^2 + (s_P - s_A)^2 + 2(1 - r) \, s_P s_A \, , \qquad (1.5.21)$$

while the other leads to

$$MSPE = (\bar{P} - \bar{A})^2 + (s_P - rs_A)^2 + (1 - r^2) \, s_A^2 \, . \qquad (1.5.22)$$

It is easy to verify that each of these equations is equivalent to (1.5.20). In the first decomposition, Theil attributes the term $(s_P - s_A)^2$ to errors of unequal variation, since this term is zero if and only if the predicted variance equals the observed variance. The final term is zero only when $r = 1$, hence we may say that a non-zero value comes about through errors of incomplete covariation. Thus the breakdown given by (1.5.21) may be expressed in words as

MSPE = errors in central tendency
+ errors due to unequal variation
+ errors due to incomplete covariation. (1.5.23)

This decomposition is completely symmetric in $\boldsymbol{P}$ and $\boldsymbol{A}$, which is not true of the second decomposition expressed in (1.5.22). This equation may be put into words as

$$\begin{aligned} MSPE &= \text{errors in central tendency} \\ &\quad + \text{errors due to regression} \\ &\quad + \text{errors due to disturbances.} \end{aligned} \tag{1.5.24}$$

The reason for this terminology is because the final term in (1.5.22) is the variation in $\boldsymbol{A}$ which is not accounted for by a least square regression of $\boldsymbol{A}$ on $\boldsymbol{P}$ — it is the 'unexplained variance' and represents the portion of MSPE which cannot be eliminated by linear corrections of the predictions. The penultimate term of (1.5.22) on the other hand is

$$s_P^2\left(1 - \frac{rs_A}{s_P}\right)^2,$$

which measures the deviation of the least squares regression coefficient (rs_A/s_P) from one, the value it would have been if the predictions were completely accurate. THEIL (1966, p. 20) gives a very simple numerical example which may be used to illustrate the two decompositions. This is based on a ten-period prediction in which the forecasts (P_i) and realizations (A_i) were as follows

i	P_i	A_i
1	5	10
2	−2	2
3	−4	−7
4	0	4
5	1	−3
6	4	6
7	7	4
8	−2	−4
9	−2	−1
10	2	3.

The MSPE for this data is

$$\tfrac{1}{10}\left[(5-10)^2 + (-2-2)^2 + \cdots + (2-3)^2\right] = 10.1\,,$$

and the decompositions corresponding to (1.5.23) and (1.5.24) respectively are

$$10.1 = 0.250 + 2.168 + 7.682 \tag{1.5.25}$$

and

$$10.1 = 0.250 + 0.115 + 9.375\,. \tag{1.5.26}$$

These may be standardized by dividing through by 10.1 to give what THEIL calls inequality proportions. In the first case these are

$$\begin{aligned} 1 &= 0.025 + 0.215 + 0.761 \\ &= u^M + u^S + u^C \,. \end{aligned} \tag{1.5.27}$$

THEIL calls u^M the bias proportion, u^S the variance proportion, and u^C the covariance proportion. In this case u^M is by far the smallest of these, reflecting the fact that the mean prediction is very close to the mean realization. The other two proportions are in a ratio of approximately 1:3, which means that in terms of this decomposition, errors due to incomplete covariation are about three times as important as those due to unequal variation.

In a similar manner, (1.5.26) may be standardized, which gives the decomposition

$$\begin{aligned} 1 &= 0.025 + 0.011 + 0.964 \\ &= u^M + u^R + u^D \,. \end{aligned}$$

u^M is as above, while u^R and u^D are respectively the regression proportion (because it deals with the deviation of the regression slope from one), and the disturbance proportion. In this case u^D dominates the decomposition. This means that with this data even the optimal linear correction of the predictions will leave 96.4% of the MSPE unaffected (THEIL 1966, p. 34).

THEIL gives an interesting geometric interpretation of his inequality measures, and also presents the useful notion of the prediction-realization diagram (THEIL 1966, pp. 37, 22). For further information on this ingenious method, the reader is referred to THEIL's original work.

1.6 Summary

This chapter has introduced the basic ideas pertaining to linear models, estimation, and prediction. Several varieties of biased estimators have been referred to, and the works of GOLDBERGER, THEIL, and BOX and JENKINS have been summarized.

2

Improved methods in estimation

2.1 What is meant by 'improved estimation'?

Speaking heuristically an 'improved estimator' may be described, following RAO (1973, p. 332), as one which is derived in the following manner. We start with some prespecified estimator, X say, having known properties of bias, variance, etc. Next we use these properties to adapt X in some way so that Y, based on X, is 'better' than X in terms of some chosen criterion function. For instance, if mean square error (MSE) is our criterion function, then we should say that Y is an 'improvement on' or 'better than' X if

$$MSE\,[Y] < MSE\,[X]\,. \tag{2.1.1}$$

This definition may be put more formally as follows. Suppose that θ is an unknown parameter which lies in some parameter space $\boldsymbol{\Omega}$, and that X is an estimator of θ. Let

$$Y = f(X, a) \tag{2.1.2}$$

be an alternative estimator, where the element a belongs to an 'adjustment space', $\boldsymbol{A}$.

Now for any given a the properties of Y and X may be compared, for instance using quadratic loss and the MSE criterion as in (2.1.1). If Y is an 'improvement' on X in terms of this criterion then we may say that a belongs to an 'improvement region', $\boldsymbol{A}^*$. In short, $\boldsymbol{A}^*$ is the subset of $\boldsymbol{A}$ which consists of the elements a such that Y given by (2.1.2) is an 'improvement' on X. Clearly $\boldsymbol{A}^*$ depends upon several things. In particular it depends on

$$\left.\begin{array}{ll}\text{(a)} & \text{the estimator } X;\\ \text{(b)} & \text{the parameter } \theta;\\ \text{(c)} & \text{the function } f;\\ \text{(d)} & \text{the criterion used for comparison.}\end{array}\right\} \tag{2.1.3}$$

As far as (c) and (d) are concerned, this book will be comparatively unadventurous. Our criterion will be almost exclusively the mean square error risk function. This criterion has little justification in principle (except its relationship to quadratic loss) and similar results could presumably be derived for other convex loss functions. But the weight of tradition and mathematical tractability has led to a consistent if misplaced emphasis upon quadratic loss, and we are unfortunately constrained to follow in this tradition. Hence rather than talking about 'improved' methods, we should perhaps refer to 'reduced mean square error' procedures, following the terminology used by STUART (1969), BLIGHT (1971), and PERLMAN (1972). See also KENDALL and STUART (1961, Vol. 2, pp. 22 and 48).

In terms of item (2.1.3c), this book concentrates upon linear functions such as

$$Y = aX\,, \tag{2.1.4}$$

and the obvious matrix generalization of this expression. Taken together with the mean square error criterion, this function has the advantage of leading to fairly tractable mathematics, which may now be illustrated using a very simple example.

Suppose that X is an unbiased estimator of the parameter θ, and that X has variance σ^2. These conditions may be written

$$E[X] = \theta \quad \text{and} \quad Var\,[X] = \sigma^2\,,$$

or as

$$X \sim (\theta, \sigma^2)\,. \tag{2.1.5}$$

Expression (2.1.5) is simply a shorthand way of writing the previous two properties — note that nothing is assumed about normality in this expression, nor of other properties beyond the first two moments.

Now if Y is given by (2.1.4) then clearly

$$E[Y] = a\theta \quad \text{and} \quad Var\,[Y] = a^2\sigma^2\,,$$

or

$$Y \sim (a\theta, a^2\sigma^2)\,. \tag{2.1.6}$$

Let us now compare the mean square error of Y with that of X. Clearly

$$\begin{aligned} MSE\,[X] &= E[(X - \theta)^2] \\ &= Var\,[X] = \sigma^2\,. \end{aligned} \tag{2.1.7}$$

On the other hand,

$$\begin{aligned} MSE\,[Y] &= E[(Y - \theta)^2] \\ &= E[(Y - E[Y])^2] + (E[Y] - \theta)^2 \\ &= Var\,[Y] + (Bias\,[Y])^2 \\ &= a^2\sigma^2 + (a\theta - \theta)^2\,. \end{aligned} \tag{2.1.8}$$

Note the general rule that "mean square error equals variance plus bias-squared" — this will prove useful in aiding our calculations. (The bias term in (2.1.7) was zero, since by assumption X is unbiased).

Now (2.1.8) can also be written

$$MSE\,[aX] = (\sigma^2 + \theta^2)\,a^2 - 2a\theta^2 + \theta^2$$
$$= (\sigma^2 + \theta^2)\,\psi(a)$$

where

$$\psi(a) = a^2 - \frac{2\theta^2}{\sigma^2 + \theta^2}\,a + \frac{\theta^2}{\sigma^2 + \theta^2} = \frac{1}{\sigma^2 + \theta^2}\,MSE\,[aX]\,. \quad (2.1.9)$$

Since ψ is proportional to the mean square error function, it can be regarded as a standardized equivalent. Its particular advantage stems from the fact that its coefficient of a^2 is unity. This helps in subsequent factorizations. For instance ψ equals

$$\psi(a) = a^2 - \frac{2}{v^2 + 1}\,a + \frac{1}{v^2 + 1}$$

where $v = \sigma/\theta$ is the coefficient of variation of X. (We assume that θ is nonzero). Then

$$\psi(a) = \left(a - \frac{1}{v^2 + 1}\right)^2 + \frac{v^2}{(v^2 + 1)^2}\,. \quad (2.1.10)$$

The standardized MSE of X is the value of $\psi(a)$ when $a = 1$. This simplifies to

$$\psi(1) = \frac{v^2}{1 + v^2}\,.$$

The ratio between this and $\psi(a)$ is

$$\frac{\psi(1)}{\psi(a)} = \frac{v^2(v^2 + 1)}{[a(v^2 + 1) - 1]^2 + v^2}\,.$$

This equals

$$\frac{MSE\,[X]}{MSE\,[aX]} = \frac{\sigma^2}{a^2\sigma^2 + (a\theta - \theta)^2} = \frac{v^2}{a^2v^2 + (a - 1)^2} \quad (2.1.11)$$

Now Y is an improvement on X when and only when this expression exceeds one. This leads to the condition

$$\frac{1 - v^2}{1 + v^2} < a < 1\,. \quad (2.1.12)$$

This inequality is important in what follows and specifies the 'improvement region' $\boldsymbol{A}^*$ within which $MSE\,[Y]$ is less than $MSE\,[X]$. Note that all scaling factors in $\boldsymbol{A}^*$ are less than one. Hence any improvement leads to

v	0	0.1	0.2	0.3	0.4	0.5	0.6	0.7	0.8	0.9
$\frac{1-v^2}{1+v^2}$	1	0.98	0.92	0.83	0.72	0.60	0.47	0.34	0.22	0.10
v	1.0	1.1	1.2	1.3	1.4	1.5	1.6	1.7	1.8	1.9
$\frac{1-v^2}{1+v^2}$	0.0	−0.10	−0.18	−0.26	−0.32	−0.38	−0.43	−0.49	−0.53	−0.57
v	2	3	4	5	6	7	8	9	10	[illegible]
$\frac{1-v^2}{1+v^2}$	−0.60	−0.80	−0.88	−0.92	−0.95	−0.96	−0.97	−0.98	−0.98	−1.00

Figure 2.1.1 Critical values of $(1 - v^2)/(1 + v^2)$ for various values of v, giving the lower bound of the improvement region (2.1.12). Values of a^* given by (2.2.1) are half-way between one and the tabulated value.

a 'shrinkage' of X towards zero. Moreover the allowable shrinkage (the lower limit of $\boldsymbol{A}^*$) is greatest when v^2 is large, and smallest when v^2 is small. In the limit as $v \to 0$ (which corresponds to $\sigma \to 0$ or $\mu \to \infty$), the improvement region tends to disappear, and with it goes all possibility for improvement by shrinkage.

Because of the importance of (2.1.12) we give in Figure 2.1.1 the critical values which lie at the lower end of the improvement region. (Note that when v exceeds one, all values of a between zero and one lead to an improvement).

The lower limit of the improvement region depends on the value of the unknown parameter θ, since this occurs in $v = \sigma/\theta$. Such a possibility was mentioned in (2.1.3b), and will in fact continually recur as one of the central difficulties with improved procedures.

2.2 'Best' improved estimation (MIMSEEs)

A slight extension of improved estimation arises by asking what value of a minimizes the mean square error of aX. In general the value of a that does this may be labelled a^*, and the corresponding estimator ($Y^* = a^*X$) may be called 'best', in the terminology of RAO (1971, and 1973 p. 305). BIBBY (1972, p. 110) called Y^* the minimum mean square error estimator (MIMSEE). However, this terminology can be misleading since there is no claim that Y^* minimizes mean square error in general, but rather that it minimizes MSE within the class of estimators derived in the way described. Nevertheless we shall retain the acronym MIMSEE where convenient. In general, MIMSEE may depend upon certain unknown parameters, as may be seen by following up the example given in the previous section.

Differentiating (2.1.8) with respect to a and putting the result equal to zero gives

$$0 = 2a\sigma^2 + 2\theta(a\theta - \theta) = 2\theta^2(av^2 + a - 1)\,,$$

where $v = \sigma/\theta$ as in (2.1.10). This equation is satisfied when

$$a = \frac{\theta^2}{\theta^2 + \sigma^2} = (1 + v^2)^{-1} = a^*\,, \quad \text{say}\,. \tag{2.2.1}$$

Under the conditions of differentiability etc., a^* is the value of a that minimizes $MSE\,[Y]$, and $Y^* = a^*X$ is the MIMSEE. Of course the same result could have been derived directly from (2.1.10). Note that a^* falls exactly at the midpoint of the improvement region $\boldsymbol{A}^*$ defined by (2.1.12). This might perhaps have been expected from the quadratic expression (2.1.10) which defines $MSE\,[aX]$, and may be illustrated by Figure 2.2.1.

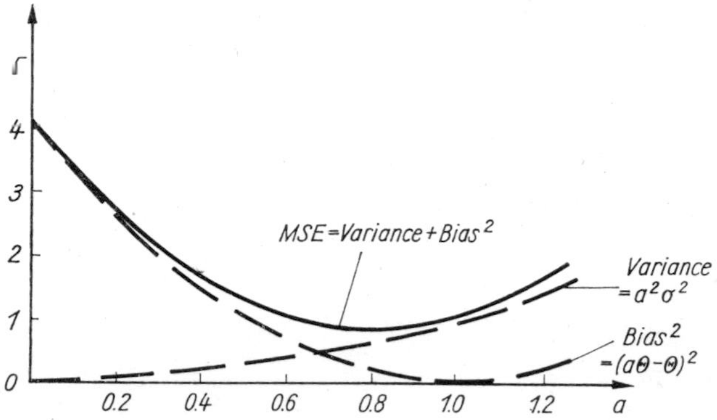

Figure 2.2.1 Graphs illustrating Variance, Bias, and MSE of aX, for values of a between 0 and 1.2 (in this example $\sigma = 1$ and $\theta = 2$, so that coefficient of variation is $v = \sigma/\theta = 1/2$, and MSE is minimized when $a = 4/5$).

Note that the variance of a^*x is

$$a^{*2}\sigma^2 = \sigma^2(1 + v^2)^{-2} ,$$

and its bias-squared is

$$(a^*\theta - \theta)^2 = \theta^2(1 - a^*)^2 = \theta^2 v^4(1 + v^2)^{-2} = \sigma^2 v^2(1 + v^2)^{-2} .$$

In other words, when $a = a^*$ the two terms in (2.1.8) have the ratio $1:v^2$. The total mean square error, their sum, is

$$MSE\ (a^*X) = \sigma^2(1 + v^2)^{-1} = \sigma^2 a^* . \tag{2.2.2}$$

That is, the estimator a^*X has a mean square error which is a^* times that of the unbiased estimator X. Again, the difference between a^* and one denotes the extent of the improvement.

The values of a^* for different values of v may be obtained from Figure 2.1.1, as half-way between one and the given value.

In this section, as in the previous one, the concepts developed seem to hinge critically upon the coefficient of variation, v. It may be thought that when we have no explicit knowledge of the value of v, these methods would therefore be useless. However that is not the case. The next section examines various types of implicit information which may be possessed about v, while Section 2.4 looks at ways of incorporating empirical information to devise a plausible iterative algorithm.

2.3 Applications

Clearly when the coefficient of variation, v, is known, the MIMSEE derived in the previous section can be calculated. However, if v is unknown, the methods discussed above involve nuisance parameters, and cannot be evaluated. We use the word 'impracticable' to describe such procedures, where the appropriate, optimal weights are unknown. This follows the original terminology used by LAPLACE in 1818, and seems less ambiguous than the translation 'infeasible' which is often used. (See STIGLER 1973, and STIGLER's contribution to the discussion of NEWBOLD and GRANGER 1974.) However, even in such situations improved estimators have their uses, especially if any of the following circumstances hold.

(a) If v is implicitly specified by the assumption that X has a particular distributional form — for instance, uniform, exponential or gamma.

(b) If there are constraints on the value of v.

(c) If prior knowledge on v exists.

These three possibilities will be examined in the following subsections, and extensions of (b) appear in Section 5.4 and Chapter 6.

2.3.1 Particular distributional forms. Suppose that X is uniformly distributed between zero and 2μ. That is, to use an obvious notation, $X \sim U(0, 2\mu)$. Then X is an unbiased estimator of μ, and its variance is $\frac{1}{3}\mu^2$. Its coefficient of variation v, is therefore $1/\sqrt{3}$. This is free of μ (i.e. it does not depend on, or is functionally independent of, μ). Hence v is 'known', in the sense that it is implicitly specified by the assumption that X has a particular distribution.

Substituting $v = 1/\sqrt{3}$ in (2.1.12) we find that the improvement region $\boldsymbol{A}^*$ for this problem is given by

$$\tfrac{1}{2} < a < 1\,. \tag{2.3.1}$$

In other words if a takes any value between $\frac{1}{2}$ and 1, then aX is an 'improvement' on X in the sense that it has lower mean square error.

The 'best' improvement takes place when $a = a^*$, given by (2.2.1). In this example $a^* = \frac{3}{4}$, the midpoint of the improvement region. Using (2.2.2) the mean square error of $Y^* = a^*X$ is $\frac{1}{4}\mu^2$, as compared with the mean square error of X which is $\frac{1}{3}\mu^2$. Hence in this case the improvement using MIMSEE has reduced the mean square error by $(1 - a^*)$ or by 25%.

More generally, if $\bar{X}$ is the mean of n independent drawings from $U(0, 2\mu)$, then

$$v^2 = \frac{\dfrac{1}{3}\mu^2/n}{\mu^2} = \frac{1}{3n}\,.$$

In this case the improvement region is

$$\frac{3n-1}{3n+1} = \frac{1 - \dfrac{1}{3n}}{1 + \dfrac{1}{3n}} < a < 1 , \tag{2.3.2}$$

and the 'best' improvement or MIMSEE occurs when $a = a^* = 3n/(3n + 1)$. Clearly when n is one this gives the value $a^* = \frac{3}{4}$ obtained in the special case considered above. As n tends to infinity, the improvement region given by (2.3.2) tends to disappear as does the scope for improvement using the methods suggested here.

Note that with the value of a^* given above, the MIMSEE is

$$a^*\bar{X} = \frac{3n}{3n+1} \times \frac{1}{n} \Sigma x_i = \frac{1}{n + \dfrac{1}{3}} \Sigma x_i \, .$$

This is just the value which the sample mean *would* have had if in addition to the n observed values there had been "one-third of an observation" which took the value zero. This interpretation of shrinkage estimators in terms of prior sample values is one which will continually recur in the examples which follow.

For instance, consider a sample of n i.i.d. $N(\mu, \sigma^2)$ random variables, with total sum of squares about the mean TSSM. The conventional unbiased estimator of σ^2 would be $\hat{\sigma}^2 = TSSM/(n - 1)$. This has mean σ^2, and variance $2\sigma^4/(n - 1)$. The squared coefficient of variation, v^2, is therefore $2/(n - 1)$, and the improvement region corresponding to (2.1.12) is

$$\frac{1 - \dfrac{2}{n-1}}{1 + \dfrac{2}{n-1}} < a < 1$$

or

$$\frac{n-3}{n+1} < a < 1 \, . \tag{2.3.3}$$

Any values of a in this region leads to an improvement, the best value being

$$a^* = \frac{n-1}{n+1} \, .$$

Hence the MIMSEE, $a^*\hat{\sigma}^2$ equals $TSSM/(n+1)$. This has a mean square error which is less that that of $\hat{\sigma}^2$, the proportional reduction being

$$1 - a^* = 2/(n+1)\,.$$

In this example we again see the possible interpretation of MIMSEE as the usual unbiased estimator based upon a sample of size $(n+2)$.
As the sample size increases, the scope for improvement diminishes, but for small samples the improvement can be quite considerable. It amounts to 67% in the special case when n equals two. For then the unbiased estimator is

$$\hat{\sigma}^2 = TSSM = (x_1 - \bar{x})^2 + (x_2 - \bar{x})^2 = \tfrac{1}{2}\,(x_1 - x_2)^2\,.$$

This has variance $2\sigma^4$ while the improved estimator is

$$a^*\hat{\sigma}^2 = \tfrac{1}{3}\,\hat{\sigma}^2 = \tfrac{1}{6}\,(x_1 - x_2)^2\,.$$

This has a downward bias of $\frac{2}{3}\,\sigma^2$, a variance of $\frac{2}{9}\,\sigma^4$, and a total mean square error of $\frac{2}{3}\,\sigma^4$, which is two thirds less than the mean square error of the unbiased estimator. (See also STUART 1969). Note also that when n equals 2 or 3, *any* value of a between zero and one lies in the improvement region specified by (2.3.3).

As a final example of the possible scope for 'improvement' using the methods advanced here, consider an exponential distribution. Its coefficient of variation is one, and hence from (2.1.12), *any* value of a between zero and one leads to an estimator which is preferable to the unbiased one. This result is noted by KENDALL and STUART (1973, vol. 2, p. 22), and of course applies to *any* distribution whose coefficient of variation is unity or greater. For instance it also applies to the problem of estimating the scale parameter of any gamma distribution whose degree of freedom parameter is known. See also Section 2.3.4.

2.3.2 Inequalities on the coefficient of variation. Suppose it is known that

$$\alpha < v < \beta$$

where α and β are two given positive numbers. Then the lower bound of the improvement region specified by (2.1.12) satisfies

$$\frac{1-\beta^2}{1+\beta^2} < \frac{1-v^2}{1+v^2} < \frac{1-\alpha^2}{1+\alpha^2}\,. \tag{2.3.4}$$

Hence the upper bound of (2.3.4) may be taken as a 'conservative indicator' of the lower end of the improvement region. Figure 2.1.1 may be used in this respect. Note that β (the upper bound on v) cannot be used to indicate bounds on the improvement region.

As an illustration of the above, suppose that σ is known to exceed $\frac{1}{2}\,\mu$. This might result either from an upper bound on μ or from a lower bound on σ. In either case $v > \frac{1}{2}$, and the 'conservative indicator' is $\frac{3}{5}$. That is, if a exceeds $\frac{3}{5}$ then it certainly exceeds the lower bound of the improvement region. One can also say in this case that the 'optimal' scaling factor, a^*, is *less than* $\frac{4}{5}$, although its precise value depends upon the extent by which σ exceeds $\frac{1}{2}\mu$.

The lower bound on v may be known a priori, or may be obtained by means of the CRAMÉR-RAO inequality, or some similar theorem (see BLIGHT 1971, and PERLMAN 1972).

Alternatively, the lower bound on v may arise through the assumption of a particular distributional form. Suppose for instance that X has a discrete uniform distribution on 1, 2, ... , N, where N is to be estimated. In these circumstances X has expectation $\frac{1}{2}(N + 1)$, and $\hat{N} = 2X - 1$ provides the conventional unbiased estimator of N. But is $\hat{N}$ the best estimator? As we shall see in the sense of MSE it certainly is it not. The squared coefficient of variation of $\hat{N}$ is

$$v^2 = \tfrac{1}{3}(N^2 - 1)/N^2 = \tfrac{1}{3}(1 - N^{-2}) .$$

Now if we have observed a particular realization of X, say x, then clearly N must be greater than or equal to x. Therefore, since v^2 is an increasing function of N,

$$v^2 \geqq \tfrac{1}{3}(1 - x^{-2}) .$$

In other words, if we now use (2.3.4) and examine the left hand inequality in (2.1.12), the lower bound of the improvement region cannot exceed

$$\frac{1 - \frac{1}{3}(1 - x^{-2})}{1 + \frac{1}{3}(1 - x^{-2})} = \frac{2x^2 + 1}{4x^2 - 1} = m\,, \quad \text{say.} \tag{2.3.5}$$

Hence we know with certainty that any a lying between m and one must give a multiplier which will lead to an 'improvement' in $\hat{N}$. That is, if

$$a = \lambda m + (1 - \lambda)\,, \tag{2.3.6}$$

where λ is any number between zero and one, than a lies in the improvement region $\boldsymbol{A}^*$. Now for any value of X, x say, the corresponding estimator $a\,\hat{N}$ is

$$\lambda m\hat{N} + (1 - \lambda)\,\hat{N} . \tag{2.3.7}$$

For instance, when $\lambda = 1$ we get the estimator $m\hat{N}$. Substituting the random variable X for the realization x in (2.3.5), this gives the estimator

$$\frac{2X^2 + 1}{4X^2 - 1}(2X - 1) .$$

Cancelling the factor $(2X - 1)$ with $4X^2 - 1 = (2X - 1)(2X + 1)$, this leads to the estimator

$$\hat{N}_1 = \frac{2X^2 + 1}{2X + 1}.$$

Note that this is asymptotically equivalent to a constant plus the maximum likelihood estimator

$$\hat{N}_2 = X,$$

which may also be obtained from (2.3.7) by choosing a suitable (random) value of λ, namely $\lambda = (2X + 1)/2(X + 1)$.

Finally, the midpoint of the known subset of the improvement region is found by putting $\lambda = \frac{1}{2}$ in (2.3.7). This gives the estimator

$$\hat{N}_3 = \frac{3X^2}{2X + 1},$$

which is asymptotically equivalent to a constant plus the estimator

$$\hat{N}_4 = \tfrac{3}{2}X.$$

It appears that all the estimators $\hat{N}_1$, $\hat{N}_2$, $\hat{N}_3$, and $\hat{N}_4$ are preferable to the conventional estimator in the sense of MSE, although for $\hat{N}_1$ and $\hat{N}_3$ at least no proof based on density functions is known to us.

The values taken by these four estimators are compared with the conventional estimator in Figure 2.3.1.

X	$\hat{N} = 2X - 1$	$\hat{N}_1 = \frac{2X^2+1}{2X+1}$	$\hat{N}_2 = X$	$\hat{N}_3 = \frac{3X^2}{2X+1}$	$\hat{N}_4 = \frac{3}{2}X$
1	1	1	1	1.50	1.50
2	3	1.80	2	2.40	3.00
3	5	2.71	3	3.86	4.50
4	7	3.67	4	5.33	6.00
5	9	4.64	5	6.82	7.50
6	11	5.62	6	8.31	9.00
7	13	6.60	7	9.80	10.50
8	15	7.59	8	11.29	12.00
9	17	8.58	9	12.79	13.50
10	19	9.57	10	14.29	15.00
20	39	19.54	20	29.27	30.00
30	59	29.52	30	44.26	45.00
40	79	39.52	40	59.26	60.00
50	99	49.51	50	74.26	75.00
100	199	99.51	100	149.25	150.00
∞	$2X - 1$	$X - \frac{1}{2}$	X	$\frac{3}{2}X - \frac{3}{4}$	$\frac{3}{2}X$

Figure 2.3.1 Comparison of various estimators for the uniform distribution. Values of the estimators $\hat{N}_1$, $\hat{N}_2$, $\hat{N}_3$, $\hat{N}_4$, compared with the conventional estimator $\hat{N} = 2X - 1$, for various values of X.

2.3.3 Possibilities for Bayesian and mixed estimation. The case where a prior distribution for v exists seems not to have been examined in the literature. However, the general procedure would be as follows. The prior distribution of v would induce a prior on the lower bound of the improvement region defined by (2.1.12). More particularly one might take the induced prior on $(1 + v^2)^{-1}$, the midpoint of the improvement region, and by basing an estimate of a^* upon this induced prior, one could obtain a Bayesian approximation to a^*X. Alternatively, of course, the observed data could be used to develop a posterior distribution for v, and the procedure outlined above could be implemented using the posterior in place of the prior.

Although this particular approach seems not have been investigated, related Bayesian work has been published by Chipman (1964) and by Lindley and Smith (1972). See also Bacon and Hausman (1974), and Zellner and Vandaele (1975).

Another possible approach would be to use a prior sample (or prior knowledge) to deduce an estimate of the value of v. This estimate could then possibly be used in association with the methods advanced here to improve the conventional estimators.

2.3.4 Estimating σ^p from a normal sample. Suppose that $X \sim N(0, \sigma^2)$, and we require to estimate σ^p where p is even. Clearly X^p has expectation $\mu_p \sigma^p$, where μ_p is the pth moment of a standard normal variable. Therefore

$$Y_p = X^p / \mu_p$$

is an unbiased estimator of σ^p. The variance of Y_p is

$$E[Y_p^2] - E([Y_p])^2 = \frac{\sigma^{2p} \mu_{2p}}{\mu_p^2} - \sigma^{2p} .$$

Hence the coefficient of variation of Y_p is v where

$$v^2 = \frac{\mu_{2p}}{\mu_p^2} - 1 . \tag{2.3.8}$$

Now μ_p in fact takes the value

$$\mu_p = \frac{p!}{(\frac{1}{2}p)!\, 2^{\frac{1}{2}p}} = 1.\, 3.\, 5.\, \ldots .\, (p-1) , \tag{2.3.9}$$

and v^2 given by (2.3.8) can easily be shown to exceed one for all even values of p. This leads to the surprising result that the improvement region defined by (2.1.12) includes *any* shrinkage factor between zero and one. The MIMSEE can also be calculated and certain values are given in Figure 2.3.2. From this it can be seen that the MIMSEE for σ^{10} is barely one-thousandth of the unbiased estimator. That is, the 'shrinkage' effect has taken away 99.9% of the conventional estimator.

p	2	4	6	8	10
μ_p	1	3	15	105	945
a^*	0.33333	0.08571	0.02164	0.00544	0.00136

Figure 2.3.2 Estimating σ^p from a normal sample — values of μ_p and a^* for small even values of p

STUART (1969) considered an extension of the above result in which there is a sample of $(n + 1)$ independent observations from $N(\mu, \sigma^2)$. An unbiased estimator of σ^p is given by

$$T_p = c_p S^p$$

where S^2 is the sum of squared deviations from the sample mean, and

$$c_p = 2^{-\frac{1}{2}p}\Gamma(\tfrac{1}{2}n)/\Gamma[\tfrac{1}{2}\ (n + p)]\ .$$

The minimum mean-square error estimator of σ^p is $a_p^* T_p$, where $a_p^* = (1 + v^2)^{-1}$. This gives the estimator

$$\frac{2^{-\frac{1}{2}p}\Gamma^2[\frac{1}{2}(n + p)]}{\Gamma[\frac{1}{2}n]\,\Gamma[\frac{1}{2}\ (n + 2p)]}\, S^p\ .$$

As STUART (1969) remarks, this also has a prior sample value interpretation, being the same multiple of S^p as is the unbiased estimator based on a sample size p members larger.

When $p = 2$ the above procedure leads to $S^2/(n + 2)$ as the estimator of the variance. Although better than both the unbiased estimator S^2 and the maximum likelihood estimator $S^2/(n + 1)$, ZACKS (1971, p. 397) shows that this estimator is itself inadmissible, being dominated by the estimator *min* $[S^2/(n + 2), 1/(n + 3)]$.

2.4 Iterative procedures

When v is unknown, and none of the methods outlined in the previous section can be used, is any alternative way forward possible?

One approach would be to estimate the nuisance parameters μ and σ, by $\tilde{\mu}$ and $\tilde{\sigma}$ say, and incorporate these estimators in (2.2.1). (We use μ now in place of θ.) When $\tilde{\mu} = X$ and $\tilde{\sigma} = S$ this gives the estimator

$$\bar{\mu} = \bar{a}X = \frac{X^3}{X^2 + S^2}\cdot \qquad (2.4.1)$$

which was investigated by THOMPSON (1968). He showed that for normal, binomial, POISSON and Gamma distributions, $\bar{\mu}$ can have larger mean square error than the conventional estimator, even when the true coefficient of variation is quite large (see KENDALL and STUART, 1973, vol. 2, p. 22).

However, Bibby (1972, pp. 110—112) argued as follows. Clearly $\bar{\mu}$ cannot dominate X uniformly, for that would conflict with the known admissibility of the conventional estimator in the case of normality. However the MSE of X is greater than the MSE of $\bar{\mu}$ for a given fixed value of $\bar{a}$ so long as $\bar{a}$ falls in the improvement region given by (2.1.12). The right-hand inequality of (2.1.12) is clearly satisfied, and the left-hand one is also satisfied if and only if

$$\frac{x^2}{s^2} > \left(\frac{1}{2}\frac{\mu^2}{\sigma^2} - 1\right), \tag{2.4.2}$$

where x and s are the observed realizations of X and S. The probability that (2.4.2) holds depends upon the precise distribution of X and S. However, if $X \sim N(\mu, \sigma^2)$ and m $S^2 \sim \sigma^2\chi^2_m$, and if they are independent, then X^2/S^2 is a non-central $F_{1,m}(v^{-2})$ variable and hence the probability that (2.4.2) holds can be calculated. (It depends only on the parameters v^2 and m.) We may conjecture that (2.4.2) 'usually' holds, in the sense that for all v and m, P[(2.4.2) holds] $> \frac{1}{2}$.

Some light may be thrown on the above conjecture by seeing what happens when σ is known (i.e. as m lends to infinity). That is instead of (2.4.2) we have

$$2x^2 > \mu^2 - \sigma^2 . \tag{2.4.3}$$

Clearly if v exceeds one, then (2.4.3) must hold, and $\bar{\mu}$ is better than the usual estimator. Thus we have demonstrated the following result.

Theorem 2.4.1 If $X \sim (\mu, \sigma^2)$ where σ is known and $\mu^2 > \sigma^2$, then X is dominated as an estimator of μ by $\bar{\mu}$ defined in (2.4.1). This holds whatever the distribution of X.

If $\mu^2 < \sigma^2$ then from (2.4.3) we know that $\hat{a}^* = X^2/(X^2 + \sigma^2)$ lies in the improvement region if

$$|X| > c(v)\,\mu \qquad \text{where } c(v) = \sqrt{\tfrac{1}{2}(1 - v^2)}\,. \tag{2.4.4}$$

Now if X is continous and the p.d.f. of X is such that

$$P(X \leq k) = \Phi\left(\frac{k - \mu}{\sigma}\right), \tag{2.4.5}$$

then

$$P(|X| > k) = 1 - \Phi\left(\frac{k - \mu}{\sigma}\right) + \Phi\left(\frac{-k - \mu}{\sigma}\right). \tag{2.4.6}$$

The probability that (2.4.4) holds may now be obtained by putting $k = c(v)\,\mu$ in (2.4.6). Writing $\sigma = v\mu$ and cancelling μ's we obtain:

$$P(2.4.4 \text{ holds}) = 1 - \Phi\left(\frac{c(v) - 1}{v}\right) + \Phi\left(\frac{-c(v) - 1}{v}\right). \tag{2.4.7}$$

If X is normal, its p.d.f. can be written in the form given by (2.4.5), where $\Phi(.)$ represents the standard normal cumulative distribution function. The numerical values of (2.4.7) for various values of v are given in Figure 2.4.1. As can be seen, the probability that $\hat{a}^*$ lies in the improvement region exceeds 0.77 for all values of v. Why then do we persist with the traditional estimator?

v	0.01	0.1	0.2	0.3	0.4	0.5	0.6	0.7	0.8	0.9	>1.0
Probability	1.0	0.998	0.937	0.860	0.811	0.782	0.770	0.776	0.799	0.842	1.00

Figure 2.4.1 Probability that a^* lies in the improvement region, when $y \sim N(\mu, \sigma^2)$ and σ is known, for various values of $v = \dfrac{\sigma}{\mu}$

The estimator $\bar{\mu}$ can also be regarded as the first stage in an iterative process. For, having calculated $\bar{\mu}$, there is no reason why we should not consider

$$\bar{\bar{\mu}} = \frac{\bar{\mu}^2}{\bar{\mu}^2 + \sigma^2} X .$$

And then having considered $\bar{\bar{\mu}}$, a third iteration is clearly possible, and so on ad infinitum.

Whittle (1962) suggests that if this process converges then its limit is

$$\tfrac{1}{2}\{x + \sqrt{(x^2 - 4\sigma^2)}\} .$$

This estimator could have a much smaller MSE than X when μ is small, but a slightly larger one (of order μ^{-4}) when μ is large. However Whittle's arithmetic appears to be faulty, and in fact the above process apparently converges to zero if $x^2 < 4\sigma^2$. Hence the iterative reductio ad infinitum is also a reductio ad absurdum, and appears not to be recommended. (See however Hemmerle 1975, who claims to give an explicit solution for the generalized ridge procedure, to which the above method is related.)

2.5 The use of nonstandard loss functions

2.5.1 Weighted mean square error. In many practical circumstances we may wish to weight the bias and standard deviation unequally. This leads to the weighted mean square error (WMSE) loss function,

$$WMSE = \lambda\,(bias)^2 + (2 - \lambda)\ variance .$$

Clearly when λ is one, WMSE is the same as MSE. In general, λ may take any value between zero and two. Extending (2.1.8) we may put

$$WMSE\,(aX) = (2-\lambda)\,a^2\sigma^2 + \lambda(a\mu - \mu)^2\,. \tag{2.5.1}$$

This is less than $WMSE\,(X)$ when

$$1 < \frac{WMSE\,(X)}{WMSE\,(aX)} = \frac{(2-\lambda)\,v^2}{(2-\lambda)\,a^2v^2 + \lambda(a-1)^2}\,.$$

This inequality may also be written

$$\frac{1-u^2}{1+u^2} < a < 1 \qquad \text{where } u^2 = (2-\lambda)\,v^2/\lambda\,. \tag{2.5.2}$$

This generalization of (2.1.12) involves simply substituting u^2 for v^2 whereever it occurs, and the rest of the work in the previous sections follows through more or less unchanged.

2.5.2 Minimum L_p norm estimation. This section examines the effect of using the L_p norm

$$L_p = |x - \mu|^p\,.$$

This leads to the so-called '*mean p-th power error*' criterion, MPE, defined as

$$MPE\,(X) = E[|X - \mu|^p]\,, \tag{2.5.3}$$

where X is an estimator, and μ is the parameter being estimated. If the value of p is required explicitly we may write (2.5.3) as $MPE_p\,(X)$. Clearly the MSE criterion is just MPE_2. One might expect MPE_1 and MPE_∞ to be related to the median and mode respectively, since these are the values that minimise the L_1 and L_∞ norms for a finite set of observations. When p is even note that

$$|\hat{\mu} - \mu|^p = (\hat{\mu} - \mu)^p = \sum_{r=0,\,p} (-1)^r \binom{p}{r} \hat{\mu}^r \mu^{p-r}\,.$$

Therefore if $\hat{\mu} = aX$ where $E[X^r] = \alpha_r$, then

$$MPE\,(\hat{\mu}) = E[|\hat{\mu} - \mu|^p] = \sum_r (-1)^r \binom{p}{r} a^r \alpha_r \mu^{p-r} = \sum_r a^r \beta_r \tag{2.5.4}$$

where

$$\beta_r = (-1)^r \binom{p}{r} \mu^{p-r} \alpha_r\,.$$

In the special case where $p = 2$ we have

$$\beta_0 = \mu^2 \alpha_0 = \mu^2$$
$$\beta_1 = -2\mu\alpha_1 = -\,2\mu^2$$

and

$$\beta_2 = \alpha_2 = \mu^2 + \sigma^2 .$$

Therefore, writing θ in place of μ, (2.1.8) is indeed a special case of (2.5.4).

The problem of minimizing (2.5.4) with respect to a is a complex one in general, but may be performed (if in no other way) by differentiating and equating the resulting polynomial to zero. (See also page 43.)

2.5.3 Some comments on mean square error (MSE). Many of the results presented in this book generalize to nonstandard convex loss functions. However, some loss functions do not lead to improvements of the type suggested here. For instance, the loss functions

$$M_1(Y) = \frac{MSE\,(Y)}{E[Y^2]} ,$$

and

$$M_2(Y) = \frac{MSE\,(Y)}{(E[Y])^2}$$

are invariant with respect to the value of a if $Y = aX$. Therefore neither of these functions allows for an improvement over the conventional unbiased estimator in the manner described here.

This observation leads one to question whether the techniques for improvement proposed in this book have in any sense a 'natural' justification, or whether they are perhaps just figments derived from the use of one particular loss function.

2.6 Multivariate extensions

The technique of improved estimation can also be extended to the multivariate case. Suppose that $\boldsymbol{X} \sim (\boldsymbol{\mu}, \boldsymbol{\Sigma})$, and that the vector $\boldsymbol{\theta}$ is some function of the elements of $\boldsymbol{\mu}$. In general we assume that $\boldsymbol{X}$ and $\boldsymbol{\mu}$ are $(p \times 1)$, while $\boldsymbol{\theta}$ is $(q \times 1)$. Sections 2.1 and 2.2 looked at the special case $p = q = 1$, and where $\boldsymbol{\theta}$ equalled $\boldsymbol{\mu}$.

The multivariate mean square error loss function of an estimator $\hat{\boldsymbol{\theta}}$ is defined as the $(q \times q)$ matrix

$$MSE\,(\hat{\boldsymbol{\theta}}) = E[(\hat{\boldsymbol{\theta}} - \boldsymbol{\theta})\,(\hat{\boldsymbol{\theta}} - \boldsymbol{\theta})'] . \tag{2.6.1}$$

Clearly when $\hat{\boldsymbol{\theta}} = \boldsymbol{X}$ and $\boldsymbol{\theta} = \boldsymbol{\mu}$ (i.e. when $\boldsymbol{X}$ is an unbiased estimator of $\boldsymbol{\mu}$) this matrix takes the value $\boldsymbol{\Sigma}$. If instead we take the estimator $\boldsymbol{Y} = \boldsymbol{AX}$ where $\boldsymbol{A}$ is a constant $(q \times p)$ matrix, then

$$MSE\,(\boldsymbol{Y}) = \boldsymbol{A\Sigma A}' + (\boldsymbol{A\mu} - \boldsymbol{\theta})\,(\boldsymbol{A\mu} - \boldsymbol{\theta})' . \tag{2.6.2}$$

This expression is a multivariate generalization of the formula "mean square error equals variance plus bias-squared", which has already been met in (2.1.8). Now when is $MSE\ (\boldsymbol{Y})$ less than $MSE\ (\boldsymbol{X})$, in the sense that the difference $MSE\ (\boldsymbol{Y}) - MSE\ (\boldsymbol{X})$ is non-positive definite? The condition for this is that

$$\boldsymbol{A\Sigma A}' + (\boldsymbol{A} - \boldsymbol{I})\,\boldsymbol{\mu\mu}'(\boldsymbol{A} - \boldsymbol{I})' - \boldsymbol{\Sigma} \quad \text{is non-positive definite,} \qquad (2.6.3)$$

a multivariate generalization of the improvement region given by (2.1.12).

An alternative approach would look at some scalar function of MSE, instead of the MSE matrix itself. For instance, we may use the trace or determinant function, and take $\hat{\boldsymbol{\theta}}_2$ rather than $\hat{\boldsymbol{\theta}}_1$ if

$$\left.\begin{array}{c} tr\ MSE\ (\hat{\boldsymbol{\theta}}_1) - tr\ MSE\ (\hat{\boldsymbol{\theta}}_2) \\ \text{or} \\ det\ MSE\ (\hat{\boldsymbol{\theta}}_1) - det\ MSE\ (\hat{\boldsymbol{\theta}}_2) \end{array}\right\} \qquad (2.6.4)$$

is positive. (Alternatively of course, one could look at the ratio of traces or determinants). These expressions have the distinct advantage of always giving a definite ordering on the two estimators, whereas it could that the expression given by (2.6.3) is neither negative definite nor positive definite. Reverting however to the matrix approach, the MIMSEE may be obtained by seeking a matrix $\boldsymbol{A}^*$ such that if $\boldsymbol{Y}^* = \boldsymbol{A}^*\boldsymbol{X}$ then

$$MSE\ (\boldsymbol{Y}^*) - MSE\ (\boldsymbol{Y}) \leqq \boldsymbol{0}$$

i.e. is non-positive definite for all other linear estimators $\boldsymbol{Y}$.

Such a matrix $\boldsymbol{A}^*$ is given by

$$\boldsymbol{\theta\mu}' = \boldsymbol{A}^*(\boldsymbol{\Sigma} + \boldsymbol{\mu\mu}') \qquad (2.6.5)$$

or

$$\boldsymbol{A}^* = \frac{\boldsymbol{\theta\mu}'\boldsymbol{\Sigma}^{-1}}{1 + \boldsymbol{\mu}'\boldsymbol{\Sigma}^{-1}\boldsymbol{\mu}} \qquad (2.6.6)$$

when $\boldsymbol{\Sigma}^{-1}$ exists. In fact

$$MSE\ (\boldsymbol{A}^*\boldsymbol{X}) = \frac{1}{1+\alpha}\,\boldsymbol{\theta\theta}',$$

where $\alpha = \boldsymbol{\mu}'\boldsymbol{\Sigma}^{-1}\boldsymbol{\mu}$ is a generalization of v^{-2}, the inverse square of the coefficient of variation.

Note that when $\boldsymbol{\mu}$ is being estimated, so that $\boldsymbol{\theta} = \boldsymbol{\mu}$, expression (2.6.6) simplifies to

$$\boldsymbol{A}^* = \boldsymbol{\mu\mu}'\boldsymbol{\Sigma}^{-1}(1 + \boldsymbol{\mu}'\boldsymbol{\Sigma}^{-1}\boldsymbol{\mu})^{-1} \qquad (2.6.7)$$

which is a generalization of the expression $v^{-2}(1 + v^{-2})^{-1}$ corresponding to (2.2.1). Other special cases of (2.6.6) are the following.

(a) When $q = 1$, so θ is a scalar, then $\boldsymbol{A}$ is a vector $\boldsymbol{a}'$ say. Then (2.6.6) becomes

$$\boldsymbol{a}^* = \theta(1 + \boldsymbol{\mu}'\boldsymbol{\Sigma}^{-1}\boldsymbol{\mu})^{-1}\,\boldsymbol{\Sigma}^{-1}\boldsymbol{\mu}$$

i.e. the elements of $\boldsymbol{a}^*$ are proportional for the elements of $\boldsymbol{\Sigma}^{-1}\boldsymbol{\mu}$. In the special case where the elements of $\boldsymbol{X}$ are unbiased estimators of θ, so that $\boldsymbol{\mu} = \theta\boldsymbol{1}$, the nuisance parameter θ cancels, and we get

$$\boldsymbol{a}^* = \theta^2(1 + \theta^2\boldsymbol{1}'\boldsymbol{\Sigma}^{-1}\boldsymbol{1})^{-1}\boldsymbol{\Sigma}^{-1}\boldsymbol{1} . \tag{2.6.8}$$

In other words, the optimal vector $\boldsymbol{a}^*$ has elements proportional to the row sums of $\boldsymbol{\Sigma}^{-1}$. This result is well known in forecasting theory.

(b) The corresponding univariate expression is obtained from (2.6.6) by putting $p = q = 1$. This gives

$$\frac{\mu\theta}{\sigma^2} \times \frac{\sigma^2}{\mu^2 + \sigma^2} \tag{2.6.9}$$

which suggests that the two terms θ^2 in (2.2.1) are of somewhat different status. The term in the denominator is the square of the mean of X, while that in the numerator is the product of this mean with the parameter being estimated.

(c) The so called 'heterogeneous' estimator $ax + c$ will be encountered several times later on in the book. This may be regarded as having the form $\boldsymbol{a}'\boldsymbol{X}$ where

$$\boldsymbol{X} = \begin{bmatrix} x \\ 1 \end{bmatrix}, \quad \boldsymbol{\mu} = \begin{bmatrix} \mu \\ 1 \end{bmatrix}, \quad \boldsymbol{a} = \begin{bmatrix} a \\ c \end{bmatrix} \quad \text{and} \quad \boldsymbol{\Sigma} = \begin{bmatrix} \sigma^2 & 0 \\ 0 & 0 \end{bmatrix}.$$

Note that $\boldsymbol{\Sigma}$ here is noninvertible, because the second element of $\boldsymbol{X}$ is constant and has zero variance. Hence $\boldsymbol{\Sigma}^{-1}$ in (2.6.6) does not exist, and we must seek some other solution. Equation (2.6.5) with $p = 2$ and $q = 1$ takes the form.

$$\theta[\mu \;\; 1] = [a \;\; c]\begin{bmatrix} \sigma^2 + \mu^2 & \mu \\ \mu & 1 \end{bmatrix}.$$

This gives two equations, namely

$$\theta\mu = a(\sigma^2 + \mu^2) + c\mu \tag{2.6.10}$$

and

$$\theta = a\mu + c . \tag{2.6.11}$$

Multiplying the second of these two equations by μ and subtracting it from the first we find that they are compatible only if

$$0 = a\sigma^2 .$$

Therefore $a = 0$ and, from (2.6.10), $c = \theta$. This highly impracticable estimator will be returned to as the 'optimal heterogeneous estimator' in Section 4.2.1.

2.7 Some historical comments

The notion of 'improved' estimation is not new, and was mentioned in 1818 by LAPLACE in his Deuxiéme Supplément à la Théorie Analytique des Probabilités. He noted that in general the mean $\bar{X}$ of a random sample has a greater asymptotic variance than

$$\bar{X} - a(\bar{X} - X_{med})$$

where X_{med} is the sample median. LAPLACE even calculated the 'correction' which led to the best improved estimator, and noted that "When one does not know the distribution of the errors of observation this correction is not feasible." It is interesting to speculate how the history of statistics would have differed if LAPLACE had followed up this observation a century and a half ago. For comments on this, and in particular the relationship of LAPLACE's work to the notion of sufficiency, see STIGLER (1973, p. 442).

2.8 Summary

This chapter has developed certain procedures for 'improving' unbiased estimators. Section 2.3 gave various applications of these procedures which used either certain types of prior knowledge, or else the pecularities of particular distributional forms.

3

Classical methods for linear models

3.1 Introduction

This chapter discusses the standard techniques of estimation, such as are dealt with in any econometric or linear model textbook (DHRYMES 1970, GOLDBERGER 1964, JOHNSTON 1972, RAO and MILLER 1971). Some parts of the chapter are developments of ideas already expressed by one author in BIBBY (1974, 1977) and in the discussion on ANDERSON (1976). We begin with a full discussion of bivariate scattergrams, a context which is sufficiently broad for many relatively sophisticated ideas to be introduced. (These tend to get obscured if developed immediately in their general multivariate form.) We then discuss fitting via multiple regression, in practical as well as theoretical terms. The GAUSS-MARKOV Theorem (often misleadingly cited as stating that the ordinary least squares line is 'best') is then proved, and its limitations emphasized. Sections 3.6 and 3.7 discuss the analysis of residuals, a much underutilized technique, while Section 3.8 goes into certain miscellaneous ideas. The final section argues for these techniques to be used with the utmost caution.

3.2 Bivariate scattergrams

As TILL (1973, p. 203) points out, 'fitting a straight line to a set of measurements is a common procedure in many branches of science', but unfortunately the ordinary least squares (OLS) method is often used where other techniques would be more appropriate. However, the first question is whether it is reasonable to represent data by straight lines at all. Polynomials or other curves are easily obtained by extending linear techniques, but more crucial is the fact that *no curve can give more information than the data on which it is based.* This point may appear self-evident, but the widespread emphasis on regression coefficients and such-like rather than the original scattergram suggests that it needs underlining. More provocatively, one may argue that science is the study of relationships between variables, that scattergrams represent the raw data before it is sucked through the sieve of statistical analysis, and that therefore *scatter-*

grams should form the starting point of any valid analytic procedure. In addition, perhaps all valid analyses should come full circle, and end up by comparing their predictions with the scattergram of observed results by means of residual analysis. In this way the adequacy of the model can be tested, and valuable extra information obtained by examining the deviations between the empirical observations and those predicted by the model. These deviations between theory and practice — POPPER's iterative schema incarnate — are often best viewed in terms of still further scattergrams (see Section 3.6). In short, *the scattergram is the fundamental analytic procedure of any branch of science.*

It then seems reasonable to ask, what extra assumptions are needed if we are to go beyond mere scattergrams, and attempt to fit lines of the form

$$y_i = ax_i + b \, ? \tag{3.2.1}$$

Firstly we note the assumption that each y_i depends only on x_i and not on x_j where $j \neq i$. Your income depends on *your* age, for instance, and not upon mine. This assumption is not always reasonable. For example, in a study of school-boys by DUNCAN, HALLER, and PORTES (1968), each boy's occupational aspirations depend upon various aspects of his best friend. Similarly, each boy's occupational achievement is related to that of his father. However, these situations can be incorporated in the analysis by considering 'father's occupation' or 'best friend's cvariables' as inputs (x's) affecting the boy's own y-variable.

A second important question is what would happen if there was replication. That is, if the same values of x appeared twice, would the same y value result ? If two boys had exactly the same background, and the same best friend, would their aspirations be exactly the same ? If this is the case, then the model is 'deterministic'. If not (as is more frequently the case) then the model is 'stochastic'.

Thirdly, where there is more than one x variable, for instance if

$$y_i = a_1 x_{i1} + a_2 x_{i2} + b \, ,$$

then we must consider whether the x variables interact. Note that interaction is synonymous with non-additivity, and is *not* the same thing as correlation. If x_1 and x_2 interact in their effect upon y, this means that a change in the value of x_2 affects the relationship between x_1 and y in a non-additive manner. That is, the curve relating x_1 and y when x_2 is zero is *not parallel* to the curve relating x_1 and y when x_2 is one. This is illustrated for linear and non-linear graphs by Figure 3.2.1. The four parts of this diagram could be represented in equation terms as follows:

A: $y = x_1 + x_2$

B: $y = x_1 + \frac{1}{2} x_1^2 + x_2$

C: $y = x_1 + 3x_2 - 2x_1 x_2$

D: $y = x_1 + \frac{1}{2} x_1^2 + 3x_2 - 2x_1 x_2 - x_1^2 x_2$

Alternatively:

C:
$$y = \begin{cases} x_1 & \text{when } x_2 = 0 \\ 3 - x_1 & \text{when } x_2 = 1 \end{cases}$$

D:
$$y = \begin{cases} x_1 + \frac{1}{2}x_1^2 & \text{when } x_2 = 0 \\ 3 - x_1 - \frac{1}{2}x_1^2 & \text{when } x_2 = 1. \end{cases}$$

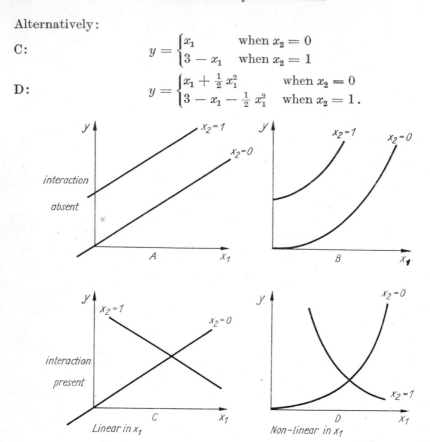

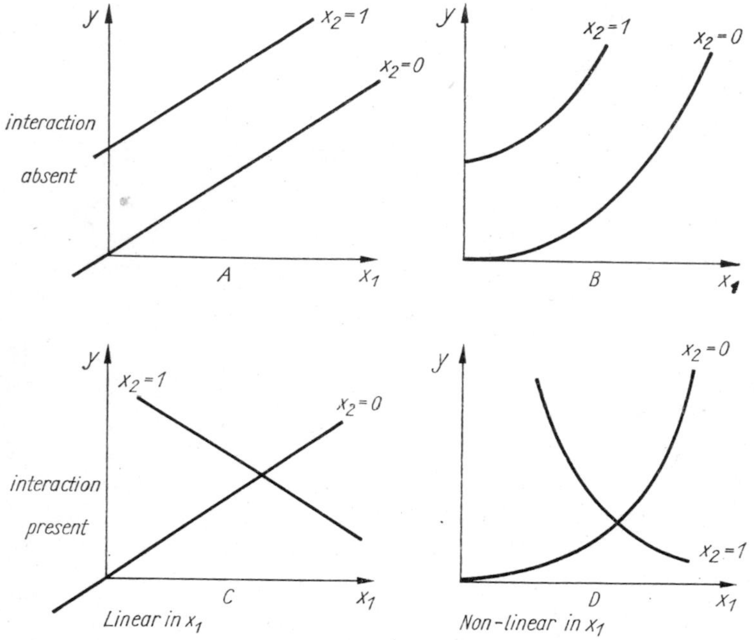

Figure 3.2.1 Illustration of interaction and linearity

These four equations also illustrate a distinction between two relationships which are linear in x_1 (A and C), and two which are nonlinear (B and D). However, by defining a new variable $x_3 = x_1^2$, both B and D could be made linear in x_1 and x_3, although in D these variables would still interact with x_2.

We return now to the situation where there is only one x variable. This may be illustrated by Figure 3.2.2, in which P_i represents a typical point with coordinates x_i and y_i. If OA is any line through the origin, then P_iQ_i represents the vertical deviation between P_i and the line OA. The OLS procedure with y as dependent variable seeks the line which minimizes $\Sigma P_iQ_i^2$. (Here and elsewhere the summation is taken over all observations). Similarly, if x is the dependent variable, OLS would minimise $\Sigma P_iR_i^2$. A third alternative defines the deviation between P_i and the line as the distance measured from P_i in a direction perpendicular to OA, viz. P_iW_i.

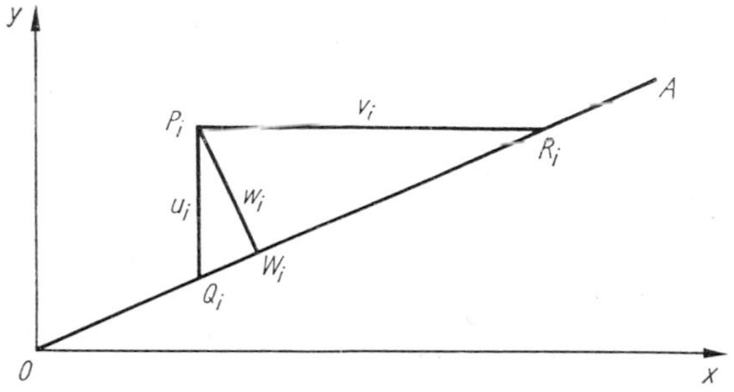

Figure 3.2.2 Diagram illustrating various line-fitting procedures.

This leads to the 'orthogonal regression procedure', ORP, which minimises $\Sigma P_i W_i^2$ (see KENDALL 1957, p. 13).

Other alternatives are possible. TILL (1973, p. 304) suggests that the sum of the triangular areas $P_iQ_iR_i$ should be minimized. This is equivalent to minimizing $\Sigma P_iQ_i \times P_iR_i$ and leads to what TILL calls the 'reduced major axis' (RMA) procedure.

We may summarize the above algebraically by noting that if the equation of the line OA is $y = ax$ then

(1) OLS with y dependent minimizes Σu_i^2 where

$$u_i = y_i - ax_i;$$

(2) OLS with x dependent minimizes Σv_i^2 where

$$v_i = x_i - a^{-1}y_i;$$

(3) ORP (orthogonal regression procedure) minimizes Σw_i^2 where

$$w_i = u_i/(1 + a^2)^{\frac{1}{2}};$$

(4) RMA (reduced major axis) minimizes $\Sigma |u_i v_i|$ where u_i and v_i are given above.

Various other procedures are considered by APPA and SMITH (1973), KIOUNTOUZIS (1973), and RICE and WHITE (1964), for instance

(5) the L_1 norm, which minimizes $\Sigma |u_i|$, and

(6) the L_∞ norm, which minimizes $\max_i |u_i|$.

These are special cases of the L_p norm of vertical deviations, which minimises $\Sigma |u_i|^p$ (see Section 2.5.2). Clearly the OLS procedure given by (1) is the L_2 norm. Also, one could equally well seek to minimize $\Sigma |v_i|^p$, or many other functions of u_i, v_i, and w_i.

Why then should OLS be so popular? RICE and WHITE (1964, p. 244) discuss the question as follows (sections in square brackets have been added by the present authors):

> "The question of choosing a criterion for smoothing and estimation has a long history. In 1799 LAPLACE [in his Méchanique Célèstre] encountered the question and proposed the use of the L_∞ norm. One was not able to perform any computation with this norm and in 1805 LEGENDRE [in the Appendix to LEGENDRE 1805] proposed the L_2 norm. At that time he demonstrated the simple computations required to determine the L_2 estimates. Since that time there has been a considerable amount of controversy about this question. Out of all the considerations of this question has been distilled the "*Principle of Least Squares*".
>
> This "principle" is normally presented (though often only implicitly) as *the* correct norm for smoothing and estimation. It is probable that this particular choice has been elevated to the position of a principle to disguise the fact that there is no single correct criterion and even if the distribution of the [disturbances] is known it is generally impossible to determine the "best" L_p norm (much less the best norm from some larger class of norms). Nevertheless, the L_2 norm is consistently presented (and usually unconditionally accepted by those who do not give serious thought to the question) and other norms are consistently ignored in the analysis of linear smoothing, estimation and regression.
>
> The principle of least squares is normally defended (if at all) on the basis of the assumption that the errors [u_i] are normally distributed. It is undoubtedly true that the L_2 norm is efficient in such a situation, probably the most efficient possible. However, we would like to refer the reader to the proposal [see GEARY 1947] that all texts on statistics should state: *Normality is a myth, there never has been, and never will be, a normal distribution.*"

However, unfortunately for this argument it is a fact that OLS can be defended even for non-normal distributions on the basis of the so-called GAUSS-MARKOV theorem, which is discussed in Section 3.4. But first we consider certain generalizations of the bivariate example given above.

3.3 Multiple regression and ordinary least squares

3.3.1 Theoretical aspects. Suppose that $\boldsymbol{y}$ is an $(n \times 1)$ vector of observations on a dependent variable, such that $E[y_i]$ is a weighted sum of the p elements of the 'explanatory vector' $\boldsymbol{x}_i$. In other words,

$$E[y_i] = \boldsymbol{x}_i'\boldsymbol{\beta} \qquad (i = 1, \dots, n)$$

where $\boldsymbol{\beta}$ is the vector of unknown weights, assumed to be the same for all i. The n equations given above can also be written in matrix form as

$$E[\boldsymbol{y}] = \boldsymbol{X}\boldsymbol{\beta}$$

where $\boldsymbol{x}_i'$ is the i-th row of the $(n \times p)$ matrix $\boldsymbol{X}$. Alternatively we may write

$$\boldsymbol{y} = \boldsymbol{X}\boldsymbol{\beta} + \boldsymbol{u} \tag{3.3.1}$$

where $\boldsymbol{u}$ is the vector of 'disturbances', representing the difference between the observed value of $\boldsymbol{y}$ and its expectation.

Although the above three expressions are equivalent, (3.3.1) is usually taken as the defining characteristic of the general linear model, upon which multiple regression is based.

Now suppose that $\boldsymbol{b}$ is any estimator of $\boldsymbol{\beta}$. This would lead to a vector of 'fitted values' given by $\boldsymbol{y}_b = \boldsymbol{X}\boldsymbol{b}$, and a 'residual' vector $\boldsymbol{u}_b$ given by

$$\boldsymbol{u}_b = \boldsymbol{y} - \boldsymbol{y}_b = \boldsymbol{y} - \boldsymbol{X}\boldsymbol{b} = \boldsymbol{X}(\boldsymbol{\beta} - \boldsymbol{b}) + \boldsymbol{u}\,. \tag{3.3.2}$$

In the last of these equations (3.3.1) has been used. The residual sum of squares (RSS) corresponding to $\boldsymbol{b}$ is

$$RSS_b = \boldsymbol{u}_b'\boldsymbol{u}_b = (\boldsymbol{y} - \boldsymbol{X}\boldsymbol{b})'\,(\boldsymbol{y} - \boldsymbol{X}\boldsymbol{b})\,.$$

Now we may ask what value of $\boldsymbol{b}$ minimizes this residual sum of squares. The answer is easily shown to be if $\boldsymbol{X}$ has full rank then RSS_b is minimized by

$$\boldsymbol{b} = (\boldsymbol{X}'\boldsymbol{X})^{-1}\boldsymbol{X}'\boldsymbol{y} = \hat{\boldsymbol{\beta}}\,, \quad \text{say}\,. \tag{3.3.3}$$

This estimator, $\hat{\boldsymbol{\beta}}$, is called the '*ordinary least squares*' (OLS) estimator. The corresponding 'fitted' value is $\hat{\boldsymbol{y}} = \boldsymbol{X}\hat{\boldsymbol{\beta}}$, and the corresponding vector of residuals is

$$\hat{\boldsymbol{u}} = \boldsymbol{y} - \hat{\boldsymbol{y}} = \boldsymbol{y} - \boldsymbol{X}\hat{\boldsymbol{\beta}} = \boldsymbol{X}(\boldsymbol{\beta} - \hat{\boldsymbol{\beta}}) + \boldsymbol{u}\,. \tag{3.3.4}$$

This equation is just a special case of (3.3.2). The corresponding residual sum of squares is

$$RSS_{\hat{\beta}} = \hat{\boldsymbol{u}}'\hat{\boldsymbol{u}} = (\boldsymbol{y} - \boldsymbol{X}\hat{\boldsymbol{\beta}})'\,(\boldsymbol{y} - \boldsymbol{X}\hat{\boldsymbol{\beta}})\,.$$

Using (3.3.3) this equals

$$\boldsymbol{y}'\boldsymbol{y} + \hat{\boldsymbol{\beta}}'\boldsymbol{X}'\boldsymbol{X}\hat{\boldsymbol{\beta}} - 2\boldsymbol{y}'\boldsymbol{X}\hat{\boldsymbol{\beta}} = \boldsymbol{y}'\boldsymbol{y} - \boldsymbol{y}'\boldsymbol{X}(\boldsymbol{X}'\boldsymbol{X})^{-1}\boldsymbol{X}'\boldsymbol{y}\,.$$

It is convenient to use the notation $\boldsymbol{G} = (\boldsymbol{X}'\boldsymbol{X})^{-1}$, $\boldsymbol{P} = \boldsymbol{X}\boldsymbol{G}\boldsymbol{X}'$, and $\boldsymbol{Q} = \boldsymbol{I} - \boldsymbol{P}$, so that

$$\hat{\boldsymbol{\beta}} = \boldsymbol{G}\boldsymbol{X}'\boldsymbol{y}\,,$$

$$\hat{\boldsymbol{y}} = \boldsymbol{P}\boldsymbol{y}\,,$$

$$\hat{\boldsymbol{u}} = \boldsymbol{Q}\boldsymbol{y}\,,$$

and

$$RSS_{\hat{\boldsymbol{\beta}}} = \boldsymbol{y}'\boldsymbol{Q}\boldsymbol{y} .$$

It will now be shown that $RSS_{\hat{\boldsymbol{\beta}}}$ is not more than $RSS_{\boldsymbol{b}}$ whatever the value of $\boldsymbol{b}$. Writing

$$\left.\begin{aligned} \boldsymbol{u}_b - \hat{\boldsymbol{u}} &= (\boldsymbol{y} - \boldsymbol{X}\boldsymbol{b}) - (\boldsymbol{y} - \boldsymbol{X}\hat{\boldsymbol{\beta}}) = \boldsymbol{X}(\hat{\boldsymbol{\beta}} - \boldsymbol{b}) , \\ &\text{we get} \\ RSS_{\boldsymbol{b}} &= \boldsymbol{u}_b'\boldsymbol{u}_b = [\hat{\boldsymbol{u}} + \boldsymbol{X}(\hat{\boldsymbol{\beta}} - \boldsymbol{b})]' \, [\hat{\boldsymbol{u}} + \boldsymbol{X}(\hat{\boldsymbol{\beta}} - \boldsymbol{b})] \\ &= \hat{\boldsymbol{u}}'\hat{\boldsymbol{u}} + (\hat{\boldsymbol{\beta}} - \boldsymbol{b})\, \boldsymbol{X}'\boldsymbol{X}(\hat{\boldsymbol{\beta}} - \boldsymbol{b}) + 2\hat{\boldsymbol{u}}'\boldsymbol{X}(\hat{\boldsymbol{\beta}} - \boldsymbol{b}) . \end{aligned}\right\} \quad (3.3.5)$$

Now the final term in this expression is zero, since

$$\hat{\boldsymbol{u}}'\boldsymbol{X} = \boldsymbol{y}'(\boldsymbol{I} - \boldsymbol{X}\boldsymbol{G}\boldsymbol{X}')\,\boldsymbol{X} = \boldsymbol{y}'\big(\boldsymbol{X} - \boldsymbol{X}(\boldsymbol{X}'\boldsymbol{X})^{-1}\,\boldsymbol{X}'\boldsymbol{X}\big) = \boldsymbol{0} .$$

Therefore from (3.3.5)

$$RSS_{\boldsymbol{b}} - RSS_{\hat{\boldsymbol{\beta}}} = (\hat{\boldsymbol{\beta}} - \boldsymbol{b})'\, \boldsymbol{X}'\boldsymbol{X}(\hat{\boldsymbol{\beta}} - \boldsymbol{b}) \geqq \boldsymbol{0} , \quad (3.3.6)$$

the inequality holding since $\boldsymbol{X}'\boldsymbol{X}$ is a non-negative definite matrix. Hence indeed $RSS_{\hat{\boldsymbol{\beta}}}$ is not more than $RSS_{\boldsymbol{b}}$ for any other estimator $\boldsymbol{b}$, and our assertion that $\hat{\boldsymbol{\beta}}$ minimizes the sum of squared residuals is justified.

What then are the properties of this OLS estimator ? Substituting (3.3.1) in (3.3.3) we see that

$$\hat{\boldsymbol{\beta}} = \boldsymbol{G}\boldsymbol{X}'\boldsymbol{y} = \boldsymbol{G}\boldsymbol{X}'(\boldsymbol{X}\boldsymbol{\beta} + \boldsymbol{u}) = \boldsymbol{\beta} + \boldsymbol{G}\boldsymbol{X}'\boldsymbol{u} . \quad (3.3.7)$$

The error of $\hat{\boldsymbol{\beta}}$ is given by

$$\boldsymbol{f} = \hat{\boldsymbol{\beta}} - \boldsymbol{\beta} = (\boldsymbol{X}'\boldsymbol{X})^{-1}\,\boldsymbol{X}'\boldsymbol{u} . \quad (3.3.8)$$

We usually assume that the disturbance term has expectation zero, i.e. that

$$E[\boldsymbol{u}] = \boldsymbol{0} . \quad (3.3.9)$$

If this is so, then (3.3.8) also has expectation zero if one of the following conditions holds (condition (c) here is a special case of condition (b)):

(a) $\boldsymbol{X}$ is fixed;
(b) $\boldsymbol{X}$ is random, but the conditional expectation of $\boldsymbol{u}$ is the same for all values of $\boldsymbol{X}$;
(c) $\boldsymbol{X}$ is random, and $\boldsymbol{u}$ is independent of $\boldsymbol{X}$. (3.3.10)

Hence we have the following result.

Theorem 3.1 If $\hat{\boldsymbol{\beta}}$ is given by (3.3.3) and $\boldsymbol{y}$ is generated according to (3.3.1), then $\hat{\boldsymbol{\beta}}$ is unbiased for $\boldsymbol{\beta}$ if and only if the expected value of (3.3.8) is zero. Sufficient conditions for this to happen are given by (3.3.9) and any one of (a), (b), or (c) in (3.3.10). (3.3.11)

The above theorem is usually stated only for the case where $\boldsymbol{X}$ is fixed. However, the extension to stochastic explanatory variables is important, especially for applications in the analysis of social surveys.

Turning now to the covariance matrix of $\hat{\boldsymbol{\beta}}$, we note from (3.3.8) that

$$\boldsymbol{V}_{\hat{\beta}} = E[\boldsymbol{f}\boldsymbol{f}'] = E[\boldsymbol{G}\boldsymbol{X}'\boldsymbol{u}\boldsymbol{u}'\boldsymbol{X}\boldsymbol{G}] \tag{3.3.12}$$

where $\boldsymbol{G} = (\boldsymbol{X}'\boldsymbol{X})^{-1}$. If $\boldsymbol{X}$ is fixed (so $\boldsymbol{G}$ is also fixed), and $E[\boldsymbol{u}\boldsymbol{u}'] = \boldsymbol{\Sigma}$ then this expression equals

$$\boldsymbol{V}_{\hat{\beta}} = \boldsymbol{G}\boldsymbol{X}'\boldsymbol{\Sigma}\boldsymbol{X}\boldsymbol{G}\,. \tag{3.3.13}$$

However, we often assume that the elements of $\boldsymbol{u}$ are uncorrelated and have the same variance, so that

$$\boldsymbol{V}_{\boldsymbol{u}} = E[\boldsymbol{u}\boldsymbol{u}'] = \boldsymbol{\Sigma} = \sigma^2\boldsymbol{I}\,.$$

In this situation, (3.3.12) simplifies whenever one of the following conditions holds:

(a) $\boldsymbol{X}$ is fixed:
(b) $\boldsymbol{X}$ is random but the conditional covariance of $\boldsymbol{u}$ is the same for all values of $\boldsymbol{X}$;
(c) $\boldsymbol{X}$ is random, and $\boldsymbol{u}$ is independent of $\boldsymbol{X}$. (3.3.14)

In case (a), (3.3.12) simplifies to $\sigma^2\boldsymbol{G}$. In the other two cases it becomes $\sigma^2 E(\boldsymbol{G})$. Hence we have the following result.

Theorem 3.2 If $\hat{\boldsymbol{\beta}}$ is given by (3.3.3), $\boldsymbol{y}$ is generated according to (3.3.1), and (3.3.9) holds, then the covariance matrix of $\hat{\boldsymbol{\beta}}$ is given by (3.3.12). In particular cases this simplifies. For instance,

if $\boldsymbol{\Sigma} = \sigma^2\boldsymbol{I}$ holds along with (3.3.14a), then the covariance of $\hat{\boldsymbol{\beta}}$ is $\sigma^2\boldsymbol{G}$;
if $\boldsymbol{\Sigma} = \sigma^2\boldsymbol{I}$ holds along with (3.3.14b), then the covariance of $\hat{\boldsymbol{\beta}}$ is $\sigma^2 E[\boldsymbol{G}]$;
if $\boldsymbol{\Sigma} = \sigma^2\boldsymbol{I}$ holds along with (3.3.14c), then the covariance of $\hat{\boldsymbol{\beta}}$ is $\sigma^2 E[\boldsymbol{G}]$. (3.3.15)

Note that in situation (a) envisaged above, the variance of $\hat{\beta}_i$ is σ^2 times the ith diagonal element of the inverse of $\boldsymbol{X}'\boldsymbol{X}$. When the columns of $\boldsymbol{X}$ are orthogonal, so that $\boldsymbol{X}'\boldsymbol{X}$ is diagonal, then the individual elements of $\hat{\boldsymbol{\beta}}$ are uncorrelated. However, this situation usually arises only where $\boldsymbol{X}$ is the design matrix of an experiment which has been specifically planned in order to induce orthogonality.

	(a) Multiple regression (general p)	Equation	(b) Bivariate without constant ($p = 1$)	(c) Bivariate with constant ($p = 2$)
Model	$\boldsymbol{y} = \boldsymbol{X\beta} + \boldsymbol{u}$	(3.3.1)	$\boldsymbol{y} = \boldsymbol{x}\beta + \boldsymbol{u}$	$\boldsymbol{u} = \boldsymbol{x}\beta + \mathbf{1}\alpha + \boldsymbol{u}$
OLS estimator	$\hat{\boldsymbol{\beta}} = (\boldsymbol{X}'\boldsymbol{X})^{-1} \boldsymbol{X}'\boldsymbol{y}$	(3.3.3)	$\hat{\beta} = (\boldsymbol{x}'\boldsymbol{x})^{-1} \boldsymbol{x}'\boldsymbol{y} = \dfrac{\Sigma x_i y_i}{\Sigma x_i^2}$	$\hat{\beta} = \dfrac{\Sigma(x_i - \bar{x})(y_i - \bar{y})}{\Sigma(x_i - \bar{x})^2}$
Variance of above	$(\boldsymbol{X}'\boldsymbol{X})^{-1} \boldsymbol{X}'\boldsymbol{\Sigma X}(\boldsymbol{X}'\boldsymbol{X})^{-1}$	(3.3.13)	$\dfrac{\boldsymbol{x}'\boldsymbol{\Sigma x}}{(\boldsymbol{x}'\boldsymbol{x})^2}$	—
Variance when $\boldsymbol{\Sigma} = \sigma^2\boldsymbol{I}$	$\sigma^2(\boldsymbol{X}'\boldsymbol{X})^{-1}$	(3.3.15)	$\sigma^2/\Sigma x_i^2$	$\sigma^2/\Sigma(x_i - \bar{x})^2$

Figure 3.3.1 A comparison of multiple regression, bivariate regression without a constant, and bivariate regression with a constant.

The most important of the above results are summarized in Figure 3.3.1, which also gives corresponding formulae for two bivariate special cases. In column (b) no constant term exists (the line passes through the origin), so that

$$\boldsymbol{X}' = [x_1 \cdots x_n] \,,$$

and p, the number of columns in $\boldsymbol{X}$, is one. In column (c) however, there is a constant, so that $p = 2$ and

$$\boldsymbol{X}' = \begin{bmatrix} x_1 \cdots x_n \\ 1 \;\cdots\; 1 \end{bmatrix}.$$

Various properties of OLS residuals may be derived from that fact that $\boldsymbol{X}'\boldsymbol{Q}$ and $\boldsymbol{P}\boldsymbol{Q}$ are both zero matrices. ($\boldsymbol{Q}$ and $\boldsymbol{P}$ are defined just after equation (3.3.4)). For instance

$$\boldsymbol{X}'\hat{\boldsymbol{u}} = \boldsymbol{X}'\boldsymbol{Q}\boldsymbol{y} = \boldsymbol{0} \,.$$

This property has already been used in deriving (3.3.6), and means that each explanatory variable has zero sum of cross-products with the residuals. Therefore, in equations with a constant term (represented by one column of $\boldsymbol{X}$ consisting solely of ones) the sum of OLS residuals is zero. This property applies for *all* sets of data. Therefore, in equations with a constant term, the observed lack of correlation between the residuals and the fitted values says nothing at all about the structure of the data — indeed, if the residuals and fitted values *were* correlated, that would indicate a computational error somewhere in the OLS calculations.

In general we reckon to have a '*good fit*' when RSS is in some sense '*small*'. (From now on RSS will be taken to mean $RSS_{\hat{\beta}}$, unless explicitly stated to the contrary). Since the size of RSS depends upon the scale of $\boldsymbol{y}$, we usually compare it with the total sum of squares about the mean,

$$TSS = \Sigma(y_i - \bar{y})^2 = \boldsymbol{y}'\boldsymbol{H}\boldsymbol{y}$$

where $\boldsymbol{H} = \boldsymbol{I} - \frac{1}{n}\boldsymbol{1}\boldsymbol{1}'$ is a centering matrix (see Glossary). This leads to the ratio

$$R^2 = 1 - \frac{RSS}{TSS} \,, \tag{3.3.16}$$

which always lies between zero and one. R^2 is low (near zero) when RSS is high (near TSS). Both these features signify a 'bad fit'. If R^2 is high (near one) then RSS is low, and we have a 'good fit'. Equation (3.3.16) is the formal definition of R^2, the square of the multiple correlation coefficient. This definition explains its common interpretation as "percentage of variance explained", since $\frac{1}{n}\,TSS$ is the variance of the dependent variable, and $\frac{1}{n}$ RSS is the 'residual variance'. (Some authors refer to R as the 'coefficient of joint determination').

An equivalent interpretation defines R as the simple correlation between the elements of $\boldsymbol{y}$ and the fitted elements of $\hat{\boldsymbol{y}}$. A third interpretation comes from the equation

$$R^2 = \Sigma r_i \hat{\beta}_i^s \,. \tag{3.3.17}$$

Here r_i is the simple correlation between the dependent variable and the ith explanatory variable, and $\hat{\beta}_i^s$ is the ith *standardized* regression coefficient, obtained when each variable has been scaled so that its variance is one. This formula has led to some misleading inferences as suggested in the next section.

3.3.2 Some practical aspects of multiple regression. In the *Plowden Report on Primary Schools,* PEAKER (1967) investigated the effect of parental and school variables upon children's academic achievement, using the technique of stepwise regression. Several varieties of this procedure exist, but PEAKER (1967, pp. 189—191) described his method as follows:

"... successive regression equations, like

$$y = b_i x_i \,, \tag{1}$$
$$y = b_i' x_i + b_j' x_j \,, \tag{2}$$
$$y = b_i'' x_i + b_j'' x_j + b_k'' x_k \tag{3}$$

etc. etc.

are produced by adding one variable at a time. At each step the computer selects the variable which, at this stage, will make the largest reduction in the remainder sum of squares. At any step it is possible that a variable previously selected may cease to be significant, and if so it is removed from the equation. When no variable remains that can make a significant reduction the process comes to an end. Thus Figure 3.3.2 was reached by the sequence of seven steps shown below:

	Standardized regression coefficients, $\hat{\beta}^s$ at step number							
Variable	1	2	3	4	5	6	7	Simple correlations r
(11)	0.415	0.365	0.315	0.307	0.263	0.254	0.252	0.415
(9)		0.356	0.318	0.295	0.252	0.241	0.251	0.408
(3)			0.236	0.204	0.197	0.190	0.200	0.370
(7)				0.194	0.203	0.263	0.248	0.312
(10)					0.150	0.148	0.148	0.355
(5)						0.144	0.150	0.067
(2)							0.102	0.076

Variable	Short description	Simple Correlation r	Standardized regression coefficient $\hat{\beta}^s$	Percentage of variance $100r\hat{\beta}^s$
Parental attitudes				
9	Aspiration for child	0.41	0.25	10.2
10	Literacy of home	0.36	0.15	5.3
11	Parental interest in school work and progress	0.42	0.25	10.5
			Parental attitude total	26.0
Home circumstances				
8	Physical amenities of home	0.18		
12	Number of dependent children	−0.13		
13	Father's occupational group	0.20		
14	Father's education	0.20		
15	Mother's education	0.16		
			Home total	0.0
School variables				
1	Teacher's sex	0.02		
2	Teacher's marital status	0.08	0.10	0.8
3	Teacher's degree of responsibility	0.37	0.20	7.4
4	Teacher's total experience	0.16		
5	Teacher's short courses	0.07	0.15	1.0
6	Teacher's long courses	0.10		
7	Teaching mark	0.31	0.25	7.7
			School total	16.9
			Grand total	42.9
			R^2	0.429
			R	0.654

Figure 3.3.2 Effect of explanatory variables upon pupil achievement (top junior boys, within school analysis). From Peaker (1967, pp. 215–216; see also p. 183).

"The first variable to be selected is, of course, the one with the largest simple correlation with the criterion. But the subsequent order of selection is not necessarily the same as the descending order of the simple correlation, as may be seen by looking down the left hand column, which shows (7) selected before (10), although the latter has a larger simple correlation. Furthermore, the last two variables to reach significance, (5) and (2), have much smaller simple correlations than several variables which are not selected at all. For example (13), which is Father's occupational group, has a simple correlation of 0.20 and there are simple correlations of 0.20 and

0.16 for (14) and (15), which are Father's education and Mother's education respectively. A variable with a fairly high simple correlation may fail to enter the regression equation because it is too highly correlated with variables that the equation already contains . . .

"It will be seen that as the process goes on the coefficients tend to settle down. Thus the changes produced by the first three steps are much greater than those produced by the last three. Variable (11), which is the pioneer, takes some hard knocks to begin with, but is pretty steady after the fifth step, while (10), which only comes in at the fifth step, is steady from the start. This illustrates a general feature of the process.

"The contributions to the assigned variation can be obtained by multiplying the regression coefficient by the simple correlation, and are as follows (in percentages):

Assigned Variation %

	Step							
Variable	1	2	3	4	5	6	7	$100\ r^2$
(11)	17.2	15.1	13.1	12.7	10.9	10.5	10.5	17.2
(9)		14.5	13.0	12.0	10.3	9.8	10.2	16.7
(3)			8.7	7.5	7.3	7.0	7.4	13.7
(7)				6.1	6.3	8.2	7.7	9.7
(10)					5.3	5.3	5.3	12.7
(5)						1.0	1.0	0.4
(2)							0.8	0.6
Total	17.2	29.6	34.8	38.3	40.1	41.8	42.9	71.0
$100\ \Sigma r^2$	17.2	33.9	47.6	57.3	70.0	70.4	71.0	

"Since each row in this table is derived from the corresponding row in the previous table by multiplying by the same quantity the new table shows the same tendency to settle down as the old one. The total of the assignable variation increases as each new variable comes in, but the increase, at first rapid, is very slight for the last three steps. This is partly because the later variables to be selected have smaller simple correlations, and partly also because they have smaller regression coefficients. The final row in the table shows what the assignable variation would be if the predictors were uncorrelated among themselves. If this were so each regression coefficient would simply be the corresponding simple correlation with the criterion. Since 42.9 is only 60 per cent of 71.0 there is a 40 per cent loss of efficiency owing to the intercorrelation of the seven predictors among themselves".

Peaker's discussion illustrates many of the points made in the previous section. But one may question his use of (3.3.17) in interpreting $r_i\hat{\beta}_i^s$ as the 'assignable variation' for variable i. Hope (1968, p. 83) calls this the 'coefficient of separate determination', and the equality given by (3.3.17) does make this terminology intuitively appealing. However it can also be deceptive. For it is perfectly conceivable that the standardized regression coefficient $\hat{\beta}_i^s$ could be positive while r_i, the correlation with the dependent variable is negative. For examples where this actually occurs see Pigeon (1967, pp. 225—228). In this case the 'assignable variation' would be negative, although the variable in question increased the variation explained by a positive amount. Therefore the concept of 'assignable variation' is somewhat misleading. Indeed the whole quest for such a measure is crying for the moon, since except for orthogonal regressors the attributable proportion explained depends not only upon the regressor itself, but also upon *the order in which the regressors are considered*. This elementary point was also overlooked by the authors of the *Coleman Report* (1966), who were criticized by BOWLES and LEVIN (1968).

It should be added that PEAKER himself no longer holds the view described above. In PEAKER (1971) and subsequent work he has discussed extensively the problems of the choice and ordering of variables, and the interpretation of regression and path analyses in the light of all the evidence. His main point is that to justify *any* interpretation the analyst must combine the immediate evidence of the study with external evidence from earlier studies and from common observation and daily experience in schools. I am most grateful to Mr. PEAKER for helpful correspondence on this point.

Some other potentially misleading aspects of multiple regression are dealt with in an excellent article by GORDON (1968). His first example illustrates how misleading regression coefficients can be if interpreted as a measure of explanatory power. For instance, Figure 3.3.3 gives several hypothetical correlation matrices. Matrix A shows five variables in two subsets having two and three members respectively. Any two variables in the same subset have a correlation of 0.8, while variables in different subsets have correlation 0.2. For instance, one subset might consist of various measures say of social class, while the other subset might contain different intelligence indices. Note that all five variables are equally useful in explaining variation in the dependent variable ($r_{yi} = 0.6$). However this equality is not reflected by the regression coefficients, which equal 0.19 in one subset and 0.27 in the other. Why is this ? The only difference between the subsets is that one contains more variables than the other — GORDON calls this the phenomenon of "differential repetitiveness", and shows (p. 597) that with a sample of size 100, the t-values are 2.70 and 1.71 respectively, reflecting significance at the 99 per cent level for the group with two members, and complete insignificance for the group with three. A similar result occurs in matrix B, which includes a third subset containing a single variable. Its correlation with the dependent variable is the same as all the others, but its regression coefficient turns out to be almost twice as large, purely by virtue of the fact that this variable is not repetitive with any other.

GORDON (1968, p. 598) cites an example of differential repetitiveness in action, where four measures of socioeconomic status and two measures of anomie were used in explaining some aspect of criminality. Within-set

Example	Correlations between Independent Variables								r_{yi}	b_{yi}	r	t	$p<$
Matrix A:		0.8	0.8	0.2	0.2				0.6	0.19	0.11	1.71	N.S.
Subsets of 3 and 2	0.8		0.8	0.2	0.2				0.6	0.19	0.11	1.71	N.S
	0.8	0.8		0.2	0.2				0.6	0.19	0.11	1.71	N.S.
	0.2	0.2	0.2		0.8				0.6	0.27	0.10	2.70	0.01
	0.2	0.2	0.2	0.8					0.6	0.27	0.10	2.70	0.01
Matrix B:		0.8	0.8	0.2	0.2	0.2			0.6	0.16	0.08	1.95	N.S.
Subsets of 3, 2,	0.8		0.8	0.2	0.2	0.2			0.6	0.16	0.08	1.95	N.S.
and 1	0.8	0.8		0.2	0.2	0.2			0.6	0.16	0.08	1.95	N.S.
	0.2	0.2	0.2		0.8	0.2			0.6	0.23	0.07	3.13	0.01
	0.2	0.2	0.2	0.8		0.2			0.6	0.23	0.07	3.13	0.01
	0.2	0.2	0.2	0.2	0.2				0.6	0.41	0.05	9.03	0.001
Matrix C:		0.8	0.8	0.8	0.2	0.2	0.2	0.2	0.6	0.12	0.09	1.41	N.S.
Subsets of 4, 3,	0.8		0.8	0.8	0.2	0.2	0.2	0.2	0.6	0.12	0.09	1.41	N.S.
and 1	0.8	0.8		0.8	0.2	0.2	0.2	0.2	0.6	0.12	0.09	1.41	N.S.
	0.8	0.8	0.8		0.2	0.2	0.2	0.2	0.6	0.12	0.09	1.41	N.S.
	0.2	0.2	0.2	0.2		0.8	0.8	0.2	0.6	0.16	0.08	1.97	N.S.
	0.2	0.2	0.2	0.2	0.8		0.8	0.2	0.6	0.16	0.08	1.97	N.S.
	0.2	0.2	0.2	0.2	0.8	0.8		0.2	0.6	0.16	0.08	1.97	N.S.
	0.2	0.2	0.2	0.2	0.2	0.2	0.2		0.6	0.40	0.05	8.83	0.001

Figure 3.3.3 Differential repetitiveness (from GORDON 1968, p. 597)

Note. – Multiple correlations, when $r_{yi} = 0.6$: for Matrix A, $R = 0.815$; Matrix B, $R = 0.905$; Matrix C, $R = 0.910$. The t-tests are based on more decimal places than are shown in the table. The standard errors and t – tests are based upon a sample size of 100.

and between-set correlations were very close, thus approximating the situation represented in Figure 3.3.3, yet the regression coefficients of the socioeconomic status variables turned out to be much less than those pertaining to anomie. This difference, it may be inferred, could have been a statistical artefact resulting from the number of measures used from each substantive domain.

Variable	(a)		(b)		(c)		(d)	
	r_{yi}	b_{yi}	r_{yi}	b_{yi}	r_{yi}	b_{yi}	r_{yi}	b_{yi}
1	0.60	0.19	0.55	0.17	0.60	0.19	0.60	0.19
2	0.60	0.19	0.55	0.17	0.60	0.19	0.60	0.19
3	0.60	0.19	0.55	0.17	0.60	0.19	0.60	0.19
4	0.60	0.27	0.60	0.28	0.55	0.24	0.60	0.38
5	0.60	0.27	0.60	0.28	0.55	0.24	0.55	0.13

Figure 3.3.4 Set of correlations with the dependent variable, r_{yi}, and corresponding regression coefficient, b_{yi} (adapted from GORDON 1968, p. 599). (Column (a) corresponds to Matrix A of Figure 3.3.3.)

A further problem arises when correlation coefficients vary within each group. Figure 3.3.4 illustrates this. Column (a) corresponds to matrix A of the previous figure. Columns (b) and (c) show the effect of slightly lowering the correlations with the dependent variable first for one group and then for the other. In either case, the effect on the regression coefficients is slight. However, when just one of the correlations in the group is lowered, as in column (d), there is a pronounced effect on the regression coefficients of the members of that group. Perhaps one should expect this. "Even so", writes GORDON (1968, p. 600), "the question could be raised as to whether a mode of analysis that transforms an 11:12 relationship (in the correlations with the dependent variable) or a 5:6 relationship (in terms of variance accounted for) into a 1:3 relationship (in the regression coefficients) provides the most helpful picture of the data. In any event, our acceptance of this outcome would be sharply revised if the outcome were based on observed values that did not reflect the true parameter values — for example, if the true correlations with the dependent variable were equal, but the observed correlations were not, or if the observed correlations reversed the direction of the true difference between the absolute correlations of two predictors with the dependent variable".

In summary, this section has developed a formal theory of multiple regression, and referred to some real-life survey-based applications. However the article by GORDON underlines the ever-present need for caution in interpreting regression results.

We turn now to examine an extension of OLS, and will then see how the residuals defined in (3.3.4) can be used to indicate whether the assumptions made in fact apply.

3.4 Generalized least squares (GLS)

3.4.1 The p-dimensional case. Equation (3.3.6) showed that the OLSE given by (3.3.3) minimizes the residual sum of squares. However, we may wish to minimize some other quadratic form of the residuals. This leads to the technique known as '*generalized least squares*' (GLS), or 'AITKEN *estimation*', after its originator. One would use GLS rather than OLS in situations where the disturbance terms (elements of $\boldsymbol{u}$) do not all have the same distribution, or where they are correlated. '*Weighted least squares*' and '*lagged least squares*' are special cases of GLS — see JOHNSTON (1972), or GOLDBERGER (1964). In the present section we merely state the GLS extension of (3.3.6) (the proof is similar) and, more importantly, the so-called 'GAUSS-MARKOV' a Theorem, which says that under certain assumptions (the 'ifs' are emphasized by TUKEY 1975 and BIBBY 1976) the GLS estimator gives minimum variance unbiased estimators of the regression coefficients.

Definition The generalized least squares estimator (GLSE) of $\boldsymbol{\beta}$ based on the dispersion matrix $\sigma^2\boldsymbol{\Sigma}$ is defined as

$$\boldsymbol{b} = \tilde{\boldsymbol{\beta}}_{\Sigma} = (\boldsymbol{X}'\boldsymbol{\Sigma}^{-1}\boldsymbol{X})^{-1}\,\boldsymbol{X}'\boldsymbol{\Sigma}^{-1}\boldsymbol{y}\,. \tag{3.4.1}$$

Note that when $\boldsymbol{\Sigma} = \sigma^2\boldsymbol{I}$ this is just the OLSE defined by (3.3.3).

Theorem 3.3 If $\hat{\boldsymbol{u}}_b$ is the residual vector $\boldsymbol{y} - \boldsymbol{X}\boldsymbol{b}$, then

$$\hat{\boldsymbol{u}}_b'\boldsymbol{\Sigma}^{-1}\hat{\boldsymbol{u}}_b \leqq \hat{\boldsymbol{u}}_{b_0}'\boldsymbol{\Sigma}^{-1}\hat{\boldsymbol{u}}_{b_0} \tag{3.4.2}$$

for any other estimator $\boldsymbol{b}_0$.

Theorem 3.4 (GAUSS-MARKOV) If $\boldsymbol{y} = \boldsymbol{X}\boldsymbol{\beta} + \boldsymbol{u}$ where $\boldsymbol{u} \sim (\boldsymbol{0}, \boldsymbol{\Sigma})$, then $\boldsymbol{b}$ is unbiased and

$$\boldsymbol{V}_b \leqq \boldsymbol{V}_{b_0} \tag{3.4.3}$$

for any other linear estimator $\boldsymbol{b}_0$. In a nutshell, the GLSE is 'best linear unbiased' (BLU).

Note that the inequality in (3.4.3) is an inequality between matrices i.e.

$$\boldsymbol{A} \leqq \boldsymbol{B}$$

denotes that the difference $(\boldsymbol{B} - \boldsymbol{A})$ is non-negative definite. Alternatively it implies that

$$\boldsymbol{c}'\boldsymbol{A}\boldsymbol{c} \leqq \boldsymbol{c}'\boldsymbol{B}\boldsymbol{c}$$

for all commensurable vectors $\boldsymbol{c}$. Applying this latter interpretation to the GAUSS-MARKOV Theorem we might also deduce the following result from (3.4.3).

Theorem 3.5 Under the conditions given in Theorem 3.4

$$Var\,[\boldsymbol{c}'\boldsymbol{b}] \leqq Var\,[\boldsymbol{c}'\boldsymbol{b}_0]\,,$$

where $\boldsymbol{b}_0$ is any other linear estimator of $\boldsymbol{\beta}$, and $\boldsymbol{c}$ is any commensurable vector.

3.4.2 Univariate examples. Some of the ideas in the above sections may become clearer if we consider the model

$$y_i = \mu + u_i \qquad (i = 1, \ldots, n)\,. \tag{3.4.4}$$

This represents a set of observations $y_1, \ldots, y_n$, each of which has mean μ. The n equations given in (3.4.4) may also be represented by

$$\boldsymbol{y} = \boldsymbol{1}\mu + \boldsymbol{u}\,, \tag{3.4.5}$$

where $\boldsymbol{1}$ is a $(n \times 1)$ vector of ones. This clearly has the form

$$\boldsymbol{y} = \boldsymbol{X\beta} + \boldsymbol{u}\,,$$

similar to (3.3.1). In (3.4.5), $\boldsymbol{X}$ is $\boldsymbol{1}$, and the $(p \times 1)$ vector $\boldsymbol{\beta}$ is the scalar μ. Corresponding to the OLSE (3.3.3) we then have

$$\hat{\boldsymbol{\beta}} = (\boldsymbol{X}'\boldsymbol{X})^{-1}\,\boldsymbol{X}'\boldsymbol{y} = (\boldsymbol{1}'\boldsymbol{1})^{-1}\,\boldsymbol{1}'\boldsymbol{y} = \frac{1}{n} \times n\bar{y} = \bar{y}\,. \tag{3.4.6}$$

In other words, the OLSE for (3.4.5) is simply the sample mean.

The GLSE (3.4.1) can also be illustrated with respect to (3.4.5). This equals

$$\boldsymbol{b} = (\boldsymbol{X}'\boldsymbol{\Sigma}^{-1}\boldsymbol{X})^{-1}\,\boldsymbol{X}'\boldsymbol{\Sigma}^{-1}\boldsymbol{y} = \boldsymbol{m}'\boldsymbol{y}/\boldsymbol{m}'\boldsymbol{1}\,, \tag{3.4.7}$$

where $\boldsymbol{m} = \boldsymbol{\Sigma}^{-1}\boldsymbol{1}$ is the vector containing the row sums of $\boldsymbol{\Sigma}^{-1}$. Equation (3.4.7) is simply a weighted mean of the elements of $\boldsymbol{y}$, the weights being proportional to the elements of $\boldsymbol{m}$. When $\boldsymbol{\Sigma}$ is diagonal — say $\boldsymbol{\Sigma} = diag\,(\sigma_1^2, \ldots, \sigma_n^2)$ — then the weight attached to y_i is simply σ_i^{-2}. That is, the terms are weighted inversely proportional to their variance.

Further examples of the GLSE are examined in Section 3.8.1.

3.5 The GAUSS-MARKOV Theorem — statement and limitations

We shall now examine in greater detail the GAUSS-MARKOV Theorem (3.4.3) for the special case where the disturbance terms are uncorrelated and homoscedastic (have the same variance), that is where $\boldsymbol{\Sigma} = \sigma^2\boldsymbol{I}$.

The statement of the GAUSS-MARKOV Theorem is as follows. Suppose that we have a 'black box' which works according to the linear model

$$\boldsymbol{y} = \boldsymbol{X\beta} + \boldsymbol{u}\,, \tag{3.5.1}$$

where $\boldsymbol{u} \sim (\boldsymbol{0}, \sigma^2\boldsymbol{I})$. Let $\hat{\boldsymbol{\beta}} = \boldsymbol{GX'y}$ be the OLS estimator of $\boldsymbol{\beta}$, and $\boldsymbol{b}_0 = \boldsymbol{Ay}$ any other linear unbiased estimator. Then, the theorem says that $\hat{\boldsymbol{\beta}}$ has a 'smaller' covariance matrix than $\boldsymbol{b}_0$. In other words

$$Covar\,[\boldsymbol{b}_0] - Covar\,[\hat{\boldsymbol{\beta}}] = \boldsymbol{V}_{\boldsymbol{b}_0} - \boldsymbol{V}_{\hat{\boldsymbol{\beta}}}$$

is non-negative definite.

Note that this theorem is couched in terms of the covariance matrix of the *estimator* $\hat{\boldsymbol{\beta}}$, and not in terms of the residual variance, which has already been shown in (3.3.6) to be less than the variance of the residuals for any other estimator. Note also that whereas (3.3.6) compares $\hat{\boldsymbol{\beta}}$ with *any* other estimator, the GAUSS-MARKOV Theorem only compares $\hat{\boldsymbol{\beta}}$ with *linear* estimators, having the form $\boldsymbol{b} = \boldsymbol{Ay}$. Finally the GAUSS-MARKOV theorem for OLS has the restrictive condition $\boldsymbol{\Sigma} = \sigma^2\boldsymbol{I}$, which is not necessary for $\hat{\boldsymbol{\beta}}$ to have the least squares property.(see McElroy 1967).

The GAUSS-MARKOV Theorem is often summarized in the statement that OLS is '*best linear unbiased*' (BLU), or more crudely still that OLS is 'best'. However, this resumé is formulated in the most presumptive of presumptive statistical terminology and can be highly misleading.

The reasons for this can be summed up by the answers to two questions, which should be investigated before any mathematical theorem is used to justify a practical procedure. These are

(1) Are the assumptions valid ?

and

(2) Are the conclusions relevant ?

In the particular context of the GAUSS-MARKOV Theorem, these questions may usefully be subdivided as follows:

(1A) Do we know that the black box is really working according to the model we have assumed, i.e. according to (3.5.1) ? In general we do not. On the contrary, we usually know that our assumed model is a crude over simplification of reality. That is, we often knowingly commit so-called '*misspecification error*'. This can take many forms. A full discussion is provided by DRAPER and SMITH (1966, pp. 81—85) and RAO and MILLER (1971, pp. 29—40 and 60—67). DEEGAN (1974) suggested a useful typology, which is discussed in BIBBY (1977). TUKEY (1975) also has a good discussion on "the seven 'ifs'" of the GAUSS-MARKOV Theorem. But the crucial point is that any optimality property such as that specified by this theorem is critically dependent upon the assumptions which form the basis of the optimality proof.

(1B) Do the disturbances have mean zero ? If not, the model can usually be reformulated so that this condition is satisfied. This is therefore a relatively unrestrictive assumption.

(1C) Are the disturbances uncorrelated with each other? This may often not be the case. For instance, in a social survey situation, disturbances relating to the same family or other cluster of respondents may well be correlated. Similarly in time series or spatial analysis, neighbouring disturbances can be highly correlated. If the *form* of this interdependence is known, then generalized least squares can be used: see Section 3.8.1.

(1D) Do the disturbances have the same variance or, in more classical vein, are they homoscedastic? They may not be, if for instance richer people tend to have a greater variance associated with their income. Again, generalized least squares can come to our aid — see Section 3.8.1.

(1E) Are the disturbances independent of the explanatory variables? This in a way is the million dollar question. If independence does not hold, the entire basis of ordinary least squares collapses. Moreover, there is no way of testing whether this devastating property obtains in any particular case.

We now turn to question 2 above, which asks whether the conclusions of the GAUSS-MARKOV Theorem have any relevance.

(2A) Are we only interested in linear estimators? If not, other procedures such as those discussed elsewhere in this book may be relevant.

(2B) Are only unbiased estimators of interest? Again, the rest of this book has relevance here. RAO and MILLER (1971, p. 65) show that a misspecified estimator, although biased, can have lower mean square error than that based on a correctly specified equation.

(2C) Is minimal variance a suitable criterion of 'bestness'? Very often not — other elements, such as cost of sampling, or some variables being more important than others, may well be important.

To say the very least, some of the above questions throw considerable doubt upon the aptness of ordinary least squares in many practical situations. For further discussion see BIBBY (1977) or Unit 15 of Open University (1977).

3.6 The use of residuals

In statistics as in other fields, one picture is very often worth a thousand words — and more than a few equations! Therefore we now reconsider the bivariate situation without a constant in graphical form. (See Column (b) of Figure 3.3.1) These graphs depend upon the important concept of a residual, which measures the deviation between one's assumed model and the observed data (see equations 3.3.2 and 3.3.4). In a way, residual analysis can be seen as the 'refutation' stage in POPPER's paradigm of science proceeding by a series of '*conjectures and refutations*'. But that, as they say, is another story.

Given a set of observations on two variables, each pair of observations (x_i, y_i) can be visualized as a point on a scatter diagram. If $n = 10$ there are ten points. The OLS procedure based on (3.3.1) fits the line through the origin which minimizes the sum of squared residuals, measured in the direction of the dependent variable. The coefficient of this line is given for the case under consideration in column (b) of Figure 3.3.1, and corresponding to each value of x there is a 'fitted' value of y, denoted by $\hat{y}$. When $x = x_i$ the fitted value is given by

$$\hat{y}_i = \hat{\beta} x_i \,. \tag{3.6.1}$$

Here $\hat{\beta} = \Sigma\, x_i y_i / \Sigma\, x_i^2$, as specified in Figure 3.3.1. The residual $\hat{u}_i$ is the difference between the observed and fitted values of the dependent variable, that is

$$\hat{u}_i = y_i - \hat{y}_i \,. \tag{3.6.2}$$

If $\hat{u}_i$ is positive, the ith point is above the OLS line; if it is negative, the point is below. Clearly y_i is the sum of $\hat{y}_i$ and $\hat{u}_i$, as illustrated in Figure 3.2.2. The residuals can be used to test questions 1A—1E, listed in Section 3.5.1, but since these questions are based on the disturbances $\boldsymbol{u}$ rather than the residuals $\hat{\boldsymbol{u}}$, we must first express the former in terms of the latter. Using (3.3.1), (3.6.1) and (3.6.2) we find that

$$\hat{u}_i = \beta x_i + u_i - \hat{\beta} x_i = (\beta - \hat{\beta})\, x_i + u_i \,.$$

That is, u_i and $\hat{u}_i$ are not equal. In fact, $\hat{u}_i$ depends on $\hat{\beta}$ and therefore involves all the n disturbance terms, not just the ith. Moreover, the disturbances are weighted by a rather complicated function of the x's, so that even if the disturbances are uncorrelated, the residuals are not; even if the disturbances have the same variances, the residuals do not. (This becomes clearer using the matrix formulation of Section 3.3, where in fact $\hat{\boldsymbol{u}} = \boldsymbol{Q}\boldsymbol{y} = \boldsymbol{Q}\boldsymbol{u}$ — using (3.3.4) and the fact that $\boldsymbol{Q}\boldsymbol{X} = \boldsymbol{0}$).

Fortunately in most cases the approximation of $\boldsymbol{u}$ by $\hat{\boldsymbol{u}}$ becomes less of a torture to reality as the sample size increases. Therefore the points emphasized in the last paragrap hcan be ignored where large samples are concerned. We shall do this for the rest of the present section, although ignoring the difference between $\boldsymbol{u}$ and $\hat{\boldsymbol{u}}$ can introduce slight errors into what follows. Our discussion is based largely on Rao and Miller (1971, Chapter 5), Draper and Smith (1966, Chapter 5), and Sprent (1969, pp. 110—112). See also Cox and Snell (1971) and Neter and Wasserman (1974, Section 4.2).

We concentrate on the Gauss-Markov Theorem mentioned in (3.4.3), and ask whether its assumptions are satisfied. That is, do the disturbances all have mean zero and the same variance, and are they statistically independent both of each other and also of the explanatory variables? If so, then we may say that they are absolutely 'random' in the colloquial sense of the word: that is, they are completely patternless, and no transformation or other jiggery-pokery can make a pattern emerge. Furthermore, no pattern should be found between the residuals and the explanatory variables. In a word, 'random' means patternless.

What this means in terms of scatter diagrams is that if a completely random variable (u) were plotted against any other variable (x) then something like Figure 3.6.1 (a) would result. If we 'step back' from this diagram then a horizontal band like Figure 3.6.1 (b) would appear, whenever u was completely random. In contrast, if our 'step back' view resembled any of the patterns in Figure 3.6.2, the very fact that we have detected a pattern would indicate a lack of randomness. Of course the particular type of pattern is important. Figure 3.6.2 (a) would indicate that the variance of u increases as x increases; Figure 3.6.2 (b) suggests the omission of a linear term in x; 3.6.2 (c) on the other hand suggests that a quadratic term should have been included, and 3.6.2 (d) might reflect the presence of outlier. In addition, a plot of u_i against u_{i-1} should show no substantial pattern if the elements of u are random.

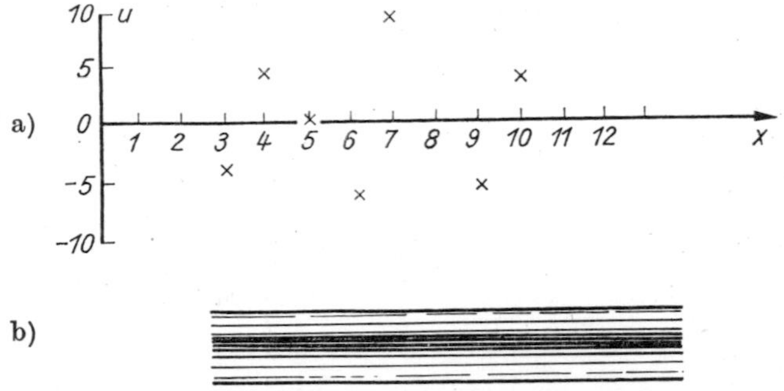

Figure 3.6.1 A typical 'all clear' residual plot, with 'step back' view (from DRAPER and SMITH, 1966, p. 89).

The above discussion was formulated in terms of a continuous variable x. But a similar analysis can be applied to categorical variables, such as sex, nationality, or social class, which may be more common in social surveys. In the simplest case, suppose that x takes just two values, 0 and 1. If u is truly random and has the same distribution for each value of x, then the 'step back' plot should be similar when $x = 0$ (say) and when $x = 1$. In contrast, the plots of Figures 3.6.3 (a) and 3.6.3 (b) suggest on the one hand that the variance of u is different in the two categories, and on the other hand that the mean of u is different. (These are the categorized analogues of parts (a) and (b) in Figure 3.6.2).

Roughly speaking, the procedures employed in the analysis of residuals are based on the rationale outlined above. However we must remember

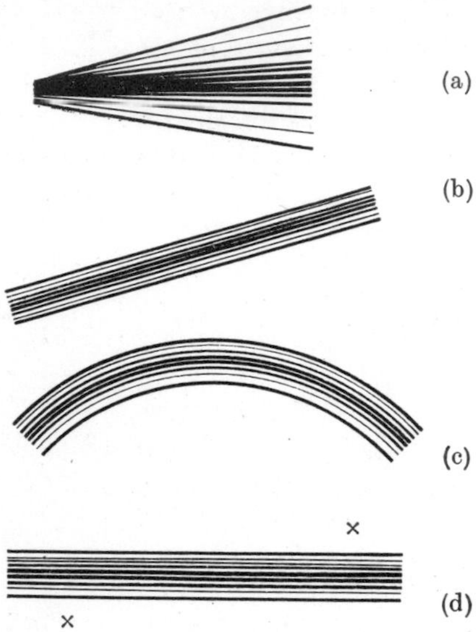

Fig. 3.6.2

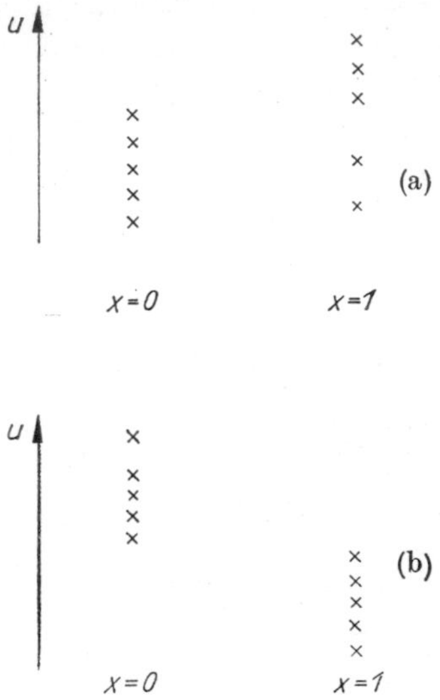

Fig. 3.6.3

Figure 3.6.2. Typical 'warning' residual plots (see text). (a) denotes increasing variance, (b) denotes omission of linear term, (c) denotes omission of quadratic term, (d) denotes presence of outliers.

Figure 3.6.3 'Warning' residual plots when x is discrete. (a) denotes differences in variance, (b) denotes differences in mean (omission of constant term).

that $\hat{u}_i$ does not equal u_i, and to that extent we are at the mercy of the assumption that our sample size is large enough to ignore this distinction.

We now return to the problem of answering the questions posed in Section 3.5. Unfortunately, they are interrelated, and one can't answer any one question without answering them all. To use a metaphor suggested by J. Johnston, we are in the position of a doctor who can correctly diagnose flu, or a broken leg, but whose diagnostic procedures cannot cope with patients suffering from both flu and a broken leg. In addition, the methods of curing these two complaints may interact, so that it is impossible at the same time both to cure the flu and to mend the broken leg. Despite this somewhat pessimistic introduction, we may proceed to discuss how residuals can help in answering some of the questions posed in Section 3.5.

(1A) Is the assumed model correct? A convenient 'catch-all' plot suggested by Draper and Smith (1966, p. 90), compares the residuals $\hat{u}_i$ with the fitted values $\hat{y}_i$. Abnormality would be indicated by plots similar to those in Figure 3.6.2. Draper and Smith state that 3.6.2 (a) would suggest the need for a transformation on the dependent variable before re-doing the regression analysis; 3.6.2 (b) suggests that a constant term has been wrongly omitted from the analysis, or some computational error has occurred; and 3.6.2 (c) indicates the need for extra explanatory variables in the model (square or cross-product term). Figure 3.6.2 (d) suggests the possibility of outliers or deviants, which may reflect no more than punching errors.

The correctness of the model could also be tested by plotting residuals against any of the other variables which are candidates for inclusion in the regression. The existence of a pattern suggest that that variable should be included. However, the absence of a pattern does not necessarily indicate the converse. As RAO and MILLER (1971, p. 115) point out, when a variable is omitted part of its influence is captured by the included variables. Therefore a plot, or correlation, between the suspected left-out variable and the residuals will not necessarily indicate any significant relation, even though in fact the variable was left out by misspecification. Therefore it is often more appropriate and economical to insert the suspected left-out variable in the equation and to estimate the coefficients afresh, rather than to try to match the residuals with the variable.

The whole problem of deciding which variables to include in a regression is very much an open one. DRAPER and SMITH surveyed the area in 1966 (Chapter 6 included '*stepwise*' and '*stagewise*' procedures), and again in 1969. The latter survey finished with the comment that "Of all the selection methods that have been suggested, none is uniquely best, and much work remains to be done". A similar conclusion might stand for the whole of the specification problem. No satisfactory purely statistical method exists for deciding which model to use. The investigator must be guided by his a priori ideas, and his knowledge of the substantive theory.

Further problems to consider are the following:

(1B) Do the disturbances have mean zero? A plot such as Figure 3.6.2 (b) would indicate that this is not the case for all values of x.

(1C) Are the disturbances uncorrelated? Again, we must have some idea of the nature of the intercorrelation in order to make progress. For instance, in social surveys the disturbances for members of the same family may be correlated. This could be tested by seeing whether the between-family variation in residuals was significantly greater than the within-family variation.

A second type of correlation between disturbances is when the labels $1, 2, \dots, n$ have a substantive meaning. This could occur if the individuals were ordered in any way, by age, income or by geographic location. The problem was first analysed in the context of time-series analysis, in which adjacent disturbances were assumed to be correlated, while non-adjacent ones are not. In this case a sensible test-statistic could be based on the OLS regression coefficient of u_i on u_{i-1}, i.e. on

$$g = \frac{\Sigma \hat{u}_i \hat{u}_{i-1}}{\Sigma \hat{u}_i^2}.$$

When the disturbances are truly independent, g has mean zero. Its variance is discussed in KENDALL (1973, p. 165), and $d = 2(1 - g)$ is the statistic proposed by DURBIN and WATSON (see RAO and MILLER, 1971, pp. 121—126 and 67—77). If the disturbances are correlated, one can use the GLS estimator outlined in Section 3.8.1.

(1D) Do the disturbances have the same variance? The possibility of unequal variance has already been mentioned with reference to Figure 3.6.2 (a). This could result from the use of aggregated data. In other cases the variance of u_i may be directly proportional to x_i^2 — this has been found in consumer surveys, where x is income and y is expenditure. In that case, say $Var\ (u_i) = \sigma^2 x_i^2$. Then instead of

$$y_i = \beta_1 + \beta_2 x_i + u_i$$

we should use

$$\frac{y_i}{x_i} = \frac{\beta_1}{x_i} + \beta_2 + \frac{u_i}{x_i}.$$

The new disturbances u_i/x_i all have variance σ^2, so as long as the other conditions are satisfied, the OLS regression of y_i/x_i upon $1/x_i$ will give 'GAUSS-MARKOV' estimates of β_1 and β_2. The coefficient of $1/x_i$ will estimate β_1, and the constant term will estimate β_2.

If instead of x_i^2, the variance of u_i is proportional to some other function, dividing through by the square root of this function will lead to equations in which the homoscedasticity condition is satisfied. For further examples see RAO and MILLER (1971, pp. 77—80 and 116—121).

(1E) Are the disturbances independent of the explanatory variables? This question pinpoints the Achilles heel of OLS regression. It is more critical than the other assumptions for two reasons. First, it is difficult to test, since we cannot observe the u's directly, and they can only be estimated under the assumption that they are independent of x. Second, if this assumption does not hold then the entire foundation of OLS estimation is destroyed, and a completely different procedure is required. Therefore where this assumption is concerned we must either simple cross our fingers and hope, or (preferably) investigate some of the procedures in Section 3.8. The reason why the collapse of this assumption is so destructive to the foundations of OLS analysis will become clear in Section 3.8. But first we examine some more examples of the analysis of residuals.

3.7 Examples of the analysis of residuals

The practical importance of residual analysis is well illustrated by an excellent series of examples due to ANSCOMBE (1973). He presented several sets of data which if analysed by standard regression procedures would all appear to present *exactly the same relationship*. Nevertheless, a careful examination of residuals indicates that the data-sets in fact display *completely different sorts of interrelationships*. This is indeed a situation where pictures — scattergrams or residual plots — speak more loudly and far more accurately than any number of words or equations. ANSCOMBE's example gives considerable support to the maxim that *a regression analysis without an examination of residuals can be positively misleading.*

ANSCOMBE's fictitious data contains four data sets, each of which has eleven (x, y) pairs. (See BIBBY (1977) or Unit 15 of Open University (1977) for full details of this example, along with the relevant tables and graphs of residuals.) It is easily confirmed that all four data-sets have the same means, variances, and correlations. These are as follows:

Mean of the x's $(\bar{x}) = 9.0$

Mean of the y's $(\bar{y}) = 7.5$

Variance of the x's $\left(\frac{1}{11} \sum (x_i - \bar{x})^2\right) = 10.0$

Variance of the y's $\left(\frac{1}{11} \sum (y_i - \hat{y})^2\right) = 3.75$

Squared correlation $(r_{xy}^2) = 0.667$

Covariance $\left(\frac{1}{11} \Sigma (x_i - \bar{x})(y_i - \bar{y})\right) = \sqrt{0.667 \times 10 \times 3.75} = 5.0.$

From these data we can calculate the value of the OLS estimator. In each case it is $5/10 = \frac{1}{2}$. Therefore since the OLS line passes through the mean point $(\bar{x}, \bar{y})$, each of the four OLS regression lines is

$$y - 7.5 = \tfrac{1}{2}(x - 9.0)$$

or

$$y = 0.5x + 3\,.$$

In addition the standard error of the estimator and the t and F-values for testing significance are identical in each of the four cases. That is, the standard analysis produced by most computer programmes would lead to identical conclusions for each of the four data sets. Nevertheless, each set demonstrates a unique form of relationship, which is only brought out by an examination of residuals (although in this extreme example the scattergrams of y against x suggest the same relationships). The fitted values and residuals can be calculated. Some peculiarities are evident from this — for instance, the residual for observation 3 of data set (c) is considerable higher than all the others for that data set. However, the full extent of the peculiarities is brought out by the plots of $\hat{u}$ and $\hat{y}$, which are shown in Figure 3.7.3. While data-set (a) shows no definite pattern (its 'step-back' picture is something like Figure 3.6.1 (b)), the same cannot be said of the other three data-sets. Set (b) has a distinct curvilinear pattern, corresponding to Figure 3.6.2 (c), while data-sets (c) and (d) each have one distinct outlier. However, if this outlier were removed the patterns displayed by (c) and (d) would be rather different. In the former case, all but one of the observations lie close to the straight line $y = 4 + 0.346x$, which is not the one yielded by the standard regression equation. Data-set (d) is rather different in that all the information about the slope of the regression line lies in a single observation — without that observation there would be no variability in x, and the slope could not be estimated. Thus the standard regression calculation in these two cases, if meaningful at all, should be accompanied by the warning that one observation has played a critical role.

Of course, the above examples are all extreme cases. But ANSCOMBE (1973, p. 20) reports a similar problem in a study of per capita school expenditure in the fifty states of the U.S.A. He reports as follows

> "... it was found that the expenditures had a satisfactory linear regression on three likely predictor variables, with multiple R^2 about 0.7 and well behaved residuals. However, one of the states, namely Alaska, was seen to have values for the predictor variables rather far removed from those of the other states, and therefore Alaska contributed rather heavily to determining the regression relation. Of course Alaska is an abnormal state, and the thought immediately occurs that perhaps Alaska should be excluded from the study. But there are other extraordinary states, Hawaii, the District of Columbia (counted here as a state), California, Florida, New York, North Dakota, ... Where does one stop*u* Rather than merely exclude Alaska, a preferable course seems to be to report the regression relation when all states are included, but add that Alaska has contributed heavily and say what happens if Alaska is omitted — the regression relation is not greatly changed, but the standard errors are increased somewhat and multiple R^2 is reduced below 0.6. We need to understand *both* the regression relation visible in all the data *and also* Alaska's special contribution to that relation."

In a survey of Accra schoolboys carried out by the present author, the residual analysis of a regression of G.C.E. performance upon Common Entrance score also proved useful. Although the regression itself had a low multiple correlation coefficient ($R^2 = 0.18$, using three explanatory variables), the residuals displayed an interesting pattern. By dividing the sample into 'over-achievers' (with positive residuals) and 'under-achievers' (with negative residuals) it was found that half the former had been taught in the 'A' stream, but only one-seventh of the latter. In the 'C' stream the proportions were reversed. Put another way, more than 80% of the A-stream over-achieved (had positive residuals), compared with only 26% of the C-stream. Hence again we see that the examination of residuals can tell us about what is happening in the real world.

3.8 Other techniques

3.8.1 Generalized least squares revisited The GAUSS-MARKOV theorem mentioned in Section 3.4.1 stated that ordinary least squares (OLS) estimators are in some sense 'best' so long as certain assumptions are satisfied. We now discuss what should be done if these assumptions are *not* satisfied. In particular, we examine the effect of an intercorrelation between the elements of the disturbance vector, $\boldsymbol{u}$. In special cases this leads to the methods of '*weighted least squares*' and of '*lagged least squares*'.

We start by defining the generalized least squares (GLS) or 'Aitken' estimator for the situation where $\boldsymbol{u}$ has mean zero and variance covariance matrix, $E(\boldsymbol{uu}') = k\boldsymbol{\Sigma}$. The GLSE is then

$$\boldsymbol{b} = (\boldsymbol{X}'\boldsymbol{\Sigma}^{-1}\boldsymbol{X})^{-1}\boldsymbol{X}'\boldsymbol{\Sigma}^{-1}\boldsymbol{y}\,, \tag{3.8.1}$$

which corresponds to the definition given in (3.4.1). In the special case where $\boldsymbol{\Sigma}$ is the identity matrix, this formula simplifies and $\boldsymbol{b}$ is just the OLS estimator given by (3.3.3). Therefore GLS is genuinely a generalization of OLS. Moreover it can be shown that the optimal properties specified by the GAUSS-MARKOV theorem accrue to the GLS estimator when the disturbance vector has covariance matrix $k\boldsymbol{\Sigma}$. (The ordinary least squares proof extends straight forwardly see GOLDBERGER 1964, p. 233). We now examine the form of $\boldsymbol{b}$ in two special cases. Firstly, suppose that the disturbances have different variances but are uncorrelated, so that the off-diagonal elements of $\boldsymbol{\Sigma}$ are zero. In other words, let $\boldsymbol{\Sigma}$ take the form

$$\boldsymbol{\Sigma} = \begin{bmatrix} \sigma_1^2 & 0 & \cdots & 0 \\ 0 & \sigma_2^2 & & 0 \\ \vdots & & \ddots & \\ 0 & 0 & \cdots & \sigma_n^2 \end{bmatrix}.$$

What does $\boldsymbol{b}$ look like in this special case? To answer this, return to the bivariate example of column (b) of Figure 3.3.1. Here the matrix $\boldsymbol{X}$ in (3.8.1) is simply a vector, and $\boldsymbol{X}'\boldsymbol{\Sigma}^{-1}\boldsymbol{X}$ is the sum $\sum x_i^2/\sigma_i^2$. Similarly, $\boldsymbol{X}'\boldsymbol{\Sigma}^{-1}\boldsymbol{y}$ is $\sum x_i y_i/\sigma_i^2$ so the GLSE in this case is

$$b = \frac{\sum \dfrac{x_i y_i}{\sigma_i^2}}{\sum \dfrac{x_i^2}{\sigma_i^2}}\,. \tag{3.8.2}$$

Again one can see how if all the σ_i's are equal, this is just the OLS estimator given in Figure 3.3.1. However when the σ_i's are not equal the terms in (3.8.2) are weighted differently, those with a higher σ_i having a lower weighting. For instance, if σ_i^2 is proportional to x_i, then b is $\sum y_i / \sum x_i = \bar{y}/\bar{x}$. If on the other hand, σ_i^2 is proportional to x_i^2, then b is $\dfrac{1}{n}\sum (y_i/x_i)$

— the mean of the ratios as opposed to the ratio of the means. (See DRAPER and SMITH 1966, p. 81, or RAO and MILLER 1971, p. 77, for further developments of these '*weighted least squares estimators*').

A second illustration of generalized least squares is given by the first-order autocorrelation process. This process is central to economic prediction, occurring frequently in time-series as well as in spatial processes in geography. In an autocorrelation process, each disturbance term is functionally related to the previous disturbance as in the following equation:

$$u_i = \varrho u_{i-1} + \varepsilon_i \,. \qquad (3.8.3)$$

The special case of independence between u_i and u_{i-1} occurs when $\varrho = 0$. In (3.8.3) the ε_i's are assumed to be independent and identically distributed, and ε_i is also independent of all the preceding u's. Under these assumptions, multiplying (3.8.3) by u_{i-1} and taking expectations gives

$$E[u_i u_{i-1}] = \varrho E[u_{i-1}^2] + E[\varepsilon_i u_{i-1}] \,,$$

or

$$covar\,(u_i, u_{i-1}) = \varrho\; var\,(u_{i-1})$$

and

$$\varrho = \frac{covar\,(u_i, u_{i-1})}{var\,(u_{i-1})} \,.$$

Therefore if the process is '*stationary*', in the sense that $var\,(u_{i-1})$ equals $var\,(u_i)$, ϱ measures the correlation between u_i and u_{i-1}. Hence it is called the *first order autocorrelation coefficient*. Similarly, the correlation between u_i and u_{i-2} is ϱ^2, that between u_i and u_{i-3} is ϱ^3, and so on. In general, the covariance matrix between the u's is a constant times

$$\boldsymbol{\Sigma} = \begin{bmatrix} 1 & \varrho & \varrho^2 & \cdots & \varrho^{n-1} \\ \varrho & 1 & \varrho & \cdots & \varrho^{n-2} \\ \varrho^2 & \varrho & 1 & \cdots & \varrho^{n-3} \\ \vdots & \vdots & \vdots & & \vdots \\ \varrho^{n-1} & \varrho^{n-2} & \varrho^{n-3} & \cdots & 1 \end{bmatrix}.$$

Now the inverse of this matrix is approximately a constant times

$$\boldsymbol{\Sigma}^{-1} = \begin{bmatrix} 1 & -\varrho & 0 & \cdots & 0 \\ 0 & 1 & -\varrho & \cdots & 0 \\ \vdots & & & \ddots & \vdots \\ 0 & \cdots & \cdots & \cdots & 1 \end{bmatrix}.$$

Let us examine what equation this will lead to for the GLS estimator in the bivariate case. In fact we find that

$$\boldsymbol{x}'\boldsymbol{\Sigma}^{-1}\boldsymbol{y} = \frac{1}{1 - 2\varrho^2} \sum (x_i - \varrho x_{i-1})\,(y_i - \varrho y_{i-1})$$

and

$$x'\Sigma^{-1}x = \frac{1}{1 - 2\varrho^2} \sum (x_i - \varrho x_{i-1})^2 .$$

Therefore from (3.8.1) the GLS estimator is

$$b = \frac{\Sigma (x_i - \varrho x_{i-1}) (y_i - \varrho y_{i-1})}{\Sigma (x_i - \varrho x_{i-1})^2} , \tag{3.8.4}$$

which of course is just the OLSE based on the first differences. This estimator is often called the *'lagged'* least squares estimator. The reason for this can be seen by considering the process which generated two adjacent observations. These are

$$y_{i-1} = \beta x_{i-1} + u_{i-1}$$

and

$$y_i = \beta x_i + u_i .$$

Since the disturbance terms here are correlated, ordinary least squares will not yield optimal estimators. But multiplying the first equation by ϱ and subtracting it from the second gives

$$y_i - \varrho y_{i-1} = \beta (x_i - \varrho x_{i-1}) + (u_i - \varrho u_{i-1}) . \tag{3.8.5}$$

Now from (3.8.3), the final term here is ε_i, and by assumption the ε's *are* independent and identically distributed. Therefore ordinary least squares on (3.8.5) *will* give an optimal estimator of β. And OLS on (3.8.5) is none other than the GLS estimator given by (3.8.4). In other words (for the proof see GOLDBERGER 1964, p. 233), in the general situation of correlated disturbances, the optimality properties of the GAUSS-MARKOV theorem accrue to the GLSE.

For other examples of generalised least squares estimators the reader is referred to HIBBS (1974), JOHNSTON (1972, pp. 208), or any econometrics textbook.

3.8.2 Multivariate regression. Although the terminology is confusing, the difference between multivariate regression and multiple regression is that the former can have more than one dependent variable. Hence as an extension of (3.3.1) we may have

$$\boldsymbol{Y} = \boldsymbol{XB} + \boldsymbol{U} , \tag{3.8.6}$$

where $\boldsymbol{Y}$ and $\boldsymbol{U}$ are $(n \times r)$ matrices, $\boldsymbol{X}$ is $(n \times q)$ and $\boldsymbol{B}$ is $(q \times r)$. Equation (3.3.1) is the special case where r equals one.

The OLS estimator of $\boldsymbol{B}$ is

$$\hat{\boldsymbol{B}} = (\boldsymbol{X}'\boldsymbol{X})^{-1} \boldsymbol{X}'\boldsymbol{Y} , \tag{3.8.7}$$

an extension of (3.3.3). In general this estimator is unbiased. Under the assumption that the rows of $\boldsymbol{U}$ are normal and independent, each with mean zero and variance $\boldsymbol{\Sigma}$, (3.8.7) is also the maximum likelihood estimator of $\boldsymbol{B}$ — see ANDERSON (1958, p. 181). The corresponding maximum likelihood estimator of $\boldsymbol{\Sigma}$ is

$$\hat{\boldsymbol{\Sigma}} = \frac{1}{n} \sum_{\alpha} (\boldsymbol{x}_\alpha - \hat{\boldsymbol{B}}\boldsymbol{y}_\alpha)(\boldsymbol{x}_\alpha - \hat{\boldsymbol{B}}\boldsymbol{y}_\alpha)' .$$

The rest of Section 3.3 is easily extended to the multivariate case, and leads on to generalizations of RSS and TSS. The multivariate extension of R^2 (equation (3.3.16)) brings us to multivariate analysis of variance, which is dealt with briefly by VAN DE GEER (1971, pp. 270–272). The distributional properties of $\hat{\boldsymbol{B}}$ and $\hat{\boldsymbol{\Sigma}}$ are discussed by ANDERSON (1958), and may be summarised by the following theorem, which uses notions of KRONECKER multiplication from Theorem A39 (see p. 169).

Theorem If $\hat{\boldsymbol{B}}$ and $\boldsymbol{Y}$ are given by (3.8.7) and (3.8.6) respectively, then

(a) $\hat{\boldsymbol{B}}$ is unbiased for $\boldsymbol{B}$ if $\boldsymbol{X}$ is fixed and

(b) if the elements of $\boldsymbol{U}$ are uncorrelated and have variances σ^2, then the variance-covariance matrix of $\hat{\boldsymbol{B}}$ is given by

$$\sigma^2\boldsymbol{I} \otimes \boldsymbol{G} \text{ where } \boldsymbol{G} = (\boldsymbol{X}'\boldsymbol{X})^{-1}. \tag{3.8.8}$$

Proof (a) Substituting (3.8.6) in (3.8.7), the error in $\hat{\boldsymbol{B}}$ is

$$\boldsymbol{F} = \hat{\boldsymbol{B}} - \boldsymbol{B} = \boldsymbol{G}\boldsymbol{X}'\boldsymbol{U} .$$

Under the given conditions, this has expectation zero. Hence these conditions are sufficient for unbiasedness (although they are not necessary).

(b) Vectorizing $\boldsymbol{F}$ we have

$$\boldsymbol{F}^v = (\boldsymbol{I} \otimes \boldsymbol{G}\boldsymbol{X}')\,\boldsymbol{U}^v .$$

Therefore under the stated conditions

$$E[\boldsymbol{F}^v\boldsymbol{F}^{v\prime}] = (\boldsymbol{I} \otimes \boldsymbol{G}\boldsymbol{X}')\,\sigma^2\boldsymbol{I}(\boldsymbol{I} \otimes \boldsymbol{X}\boldsymbol{G}) = \sigma^2\boldsymbol{I} \otimes \boldsymbol{G} .$$

When the elements of $\boldsymbol{U}$ are not all uncorrelated, other procedures may be preferable to OLS. For instance, it may be that the disturbances are correlated between time points, but not between variables. In this case, $E[\boldsymbol{U}^v\boldsymbol{U}^{v\prime}]$ takes the form

$$\begin{bmatrix} \boldsymbol{\Sigma}_1 & & \boldsymbol{0} \\ & \boldsymbol{\Sigma}_2 & \\ & & \ddots \\ \boldsymbol{0} & & \boldsymbol{\Sigma}_T \end{bmatrix}$$

where all off-diagonal matrices are zero. If $\boldsymbol{\Sigma}_1 = \boldsymbol{\Sigma}_2 = \cdots = \boldsymbol{\Sigma}_T = \boldsymbol{\Sigma}$, say, so that all variables are equally correlated, this expression is just $\boldsymbol{I} \otimes \boldsymbol{\Sigma}$.

Alternatively, we may have a correlation between variables measured at the same time point, but none between different time points. In this case, if the correlation matrix at each time point is the same (say $\boldsymbol{\Sigma}$), then $E[\mathbf{U}^v\mathbf{U}^{v\prime}] = \boldsymbol{\Sigma} \otimes \mathbf{I}$.

These observations lead to the following result.

Theorem If $\hat{\mathbf{B}}$ and $\mathbf{Y}$ are given by (3.8.7) and (3.8.6) respectively, and if $\mathbf{X}$ in (3.8.6) is fixed and $E[\mathbf{U}] = \mathbf{0}$, then

(a) If the elements of $\mathbf{U}$ are uncorrelated between variables, but all variables have the same covariance ($\boldsymbol{\Sigma}$) between time points, then the covariance matrix of $\hat{\mathbf{B}}$ is $\boldsymbol{\Sigma} \otimes \mathbf{G}$;

(b) If the elements of $\mathbf{U}$ are uncorrelated between time points, but all time points have the same covariance ($\boldsymbol{\Sigma}$) between variables, then the covariance matrix of $\hat{\mathbf{B}}$ is $\mathbf{I} \otimes \mathbf{G}\mathbf{X}'\boldsymbol{\Sigma}\mathbf{X}\mathbf{G}$. (3.8.9)

In fact the situation envisaged by part (b) of Theorem 3.8.9 is much more common than that in part (a). See also the situation of '*seemingly unrelated regressions*' treated by ZELLNER (1962).

3.8.3 Simultaneous equation models. The last section examined a system whose several dependent variables each had the same set of explanatory variables. We now add the possibility that one dependent variable can also be an explanatory variable influencing one of the other dependent variables. This situation is illustrated in Figure 3.8.1.

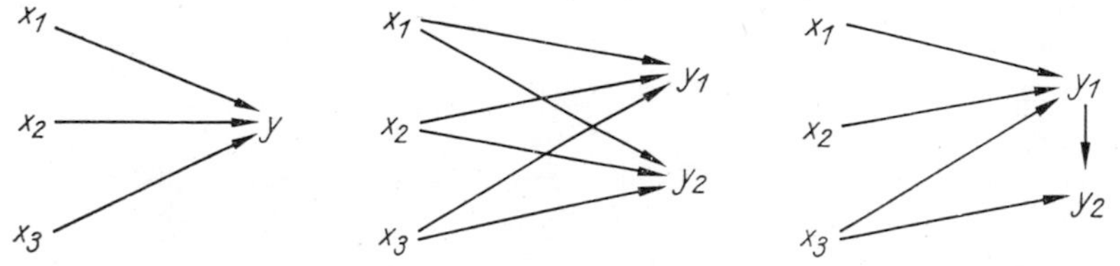

multiple regression — multivariate regression — simultaneous equations

Figure 3.8.1 Diagram representing the difference between multiple regression, multivariate regression, and simultaneous equations.

Although in any given equation one can distinguish between a dependent variable and an explanatory one, in the model as a whole this distinction need no longer exist, since variables may play different roles in different equations. Hence some new terminology is necessary. This terminology may be illustrated with an example. Suppose for example that

$$y_{i1} = x_{i1}\beta_{11} + u_{i1} \quad \text{(a)}$$

and

$$y_{i2} = x_{i1}\beta_{12} + y_{i1}\gamma_{11} + u_{i2}\,. \quad \text{(b)} \qquad (3.8.10)$$

While x_1 is the only 'explanatory' variable in the first equation, both x_1 and y_1 are explanatory in the second. However, as far as the system as a whole is concerned, we may distinguish between x_1 which is determined outside the system, and y_1 and y_2 which are not. The former variables are called '*exogenous*', while the latter variables are '*endogenous*'. Endogenous variables appear as the dependent variable in one of the equations, while exogenous variables do not. In terms of arrow diagrams, an endogenous variable has arrows pointing *to* it, while an exogenous variable only has arrows pointing *away* from it.

3.8.4 Econometric estimation. Simultaneous equations can also be brought into the formulation given by (3.8.6), using the matrix equation

$$\boldsymbol{Y} = \boldsymbol{X}\boldsymbol{C} + \boldsymbol{Y}\boldsymbol{D} + \boldsymbol{V} . \tag{3.8.11}$$

The elements of the $\boldsymbol{Y}$ matrix are observations on endogeneous variables, while the elements of $\boldsymbol{X}$ are exogenous. From this equation clearly

$$\boldsymbol{Y} = \boldsymbol{X}\boldsymbol{C}(\boldsymbol{I} - \boldsymbol{D})^{-1} + \boldsymbol{V}(\boldsymbol{I} - \boldsymbol{D})^{-1} , \tag{3.8.12}$$

which on putting $\boldsymbol{B} = \boldsymbol{C}(\boldsymbol{I} - \boldsymbol{D})^{-1}$ and $\boldsymbol{U} = \boldsymbol{V}(\boldsymbol{I} - \boldsymbol{D})^{-1}$ leads us straight back to (3.8.6). Equation (3.8.11) is called the '*structural form*' of the model, and the matrices $\boldsymbol{C}$ and $\boldsymbol{D}$ give '*structural form coefficients*'. In contrast, (3.8.12) is the '*reduced form*', and the elements of $\boldsymbol{C}(\boldsymbol{I} - \boldsymbol{D})^{-1}$ are '*reduced form coefficients*'. The latter (but not necessarily the former) can be deduced using principles of multivariate regression. For further details see DHRYMES (1970, p. 279).

3.9 Conclusion — the need for caution

We wish to end this chapter by emphasizing that the linear model can be a trap, a snare, and a delusion — although these evil effects may be partially mitigated by adhering to the cautionary words of advice given in previous sections. Despite its dangers, the GLM is very popular among practitioners. (Sociologists of knowledge could well enquire why this is so. By examining the reasons for its popularity, they might learn something about the GLM, but would certainly learn a great deal about those who use it.)

"Seek simplicity . . .," said WHITEHEAD. And in a way the linear model is extremely simple. It certainly eliminates the complexities of hardheaded thought, especially since so many computer programs exist. For the soft-headed analyst who doesn't want to think too much, an off-the-peg computer package is simplicity itself, especially if it cuts through a mass of complicated data and provides a few easily reportable coefficients. OCCAM's razor has been used to justify worse barbarities: but razors are dangerous things and should be used carefully. Moreover the simple outcomes of the general linear model can only tell half the tale, and should be viewed with scepticism. Simplicity is by its nature a travesty of truth. "Seek simplicity . . ." said Whitehead, ". . . and distrust it".

One final point which is worthy of particular comment is the way in which the use of the linear model and other sophisticated statistical techniques can change the terms of the substantive debate. We may take the example of a schoolboy survey, whose original aims were quite simple and easy to understand (see DUNCAN et al. 1968 and BIBBY 1977). These aims concerned the question of what factors affect the educational and occupational aspirations of schoolboys. This simple question nevertheless allows a considerable richness of possible interpretations and responses. It is a question that is easily understood and that many people would be prepared to have a go at answering. Yet its re-formulation in terms of a linear model changes the situation in various identifiable ways.

Firstly, this reformulation *narrows* the question by restricting its terms of reference to a particular model based upon a particular set of observed empirical data. Secondly, it *trivializes* the question by removing its substantive richness and substituting apparently meaningful yet potentially trivial questions concerning the values of unknown regression coefficients. A third effect is to *technicalize* the debate — instead of discussions concerning the nature of the social forces in action, the debate becomes concentrated on technical matters such as bias, mis-specification and optimality. One result of this technical emphasis is to *obscure* the original question for which an answer was being sought. Finally, statistical techniques tend to *expertize* the debate. That is, while the original question could be stated, understood, and answered by the ordinary literate man in the street, the new 'statistical' question is formulated in such a way that only a few experts can understand it, let alone express an opinion.

To summarize, the use of the linear models can twist the terms of the debate in several unwitting and unwelcome ways — in particular we have mentioned the effects of narrowing, trivialization, technicalization, obscurantism, and expertization. These aspects of political abuse are just as important as the areas of 'technical abuse' which are usually emphasized in textbooks of this sort. If one is to teach a responsible use of statistics by means of the 'use-abuse paradigm', it is important for these aspects of political abuse to be placed at the centre of the stage.

3.10 Summary

This chapter has discussed certain classical methods for analysing linear models. Our treatment has however differed from that conventionally adopted in the following respects.

(a) The importance of scatter diagrams has been emphasized in preference to algebraic formulations.
(b) The scientific centrality of residuals has been noted — the deviations from the pattern can be more important than the pattern itself.
(c) The treatment of the GAUSS-MARKOV theorem emphasized the assumptions more than the conclusions.
(d) The need for caution and the prospects for political misuse of statistics have been mentioned — this can be more important than technical misuse.

4

Improved estimation in linear models

4.1 Introduction

This chapter returns to the general linear model

$$y = X\beta + u\,, \qquad u \sim (0, \sigma^2 W) \tag{4.1.1}$$

specified in Section 3.3.1. Note that the covariance matrix is now called $\sigma^2 W$ instead of Σ. W is positive definite, $X(n \times p)$ is fixed, and *rank* $(X) = p$. SCHÖNFELD (1969) considered the problem of estimating simultaneously the elements of $\sigma^2 W$ along with those of β, but this chapter assumes that W is known. Hence we are concerned with the simultaneous estimation of β and σ^2 (that is $p + 1$ parameters in all).

We shall consider primarily those estimators of β which are linear in y i.e. which can be written in the form

$$\hat{\beta} = C'y + d\,. \tag{4.1.2}$$

If d is zero, $\hat{\beta}$ will be called a 'homogeneous linear estimator'. Otherwise it is 'heterogeneous'. (Note that $\hat{\beta}$ defined by (4.1.2) should not be confused with the OLS estimator defined in (3.3.3).)

As a basis for comparing estimators we shall take the quadratic loss function

$$R(\hat{\beta}) = E[(\hat{\beta} - \beta)' A(\hat{\beta} - \beta)]\,, \tag{4.1.3}$$

where A is a positive definite $(p \times p)$ matrix selected according to the relative importance placed upon the different elements of $\hat{\beta}$. Often A may be taken to be the identity matrix, in which case $R(\hat{\beta})$ represents the expected EUCLIDEAN distance between the true value β and its estimator. In general, $R(.)$ represents a MAHALONOBIS distance in the metric defined by A^{-1}.

Of course other loss functions could also be used. TOUTENBURG (1975, p. 55) considered

$$\left.\begin{aligned} &R_2(\hat{\beta}) = E[(\hat{\beta} - \beta)' a]^2 \\ \text{and}\quad &R_3(\hat{\beta}) = E[(y - X\hat{\beta})' W^{-1}(y - X\hat{\beta})] \end{aligned}\right\} \tag{4.1.4}$$

as well as $R(.)$ defined in (4.1.3). (TOUTENBURG (1975) used the notation $R_1(.)$ in place of $R(.)$.) However both $R_3(.)$ and $R_2(.)$ may be related to $R(.)$, which also has the advantage of tractability. Hence this book concentrates upon the loss function $R(.)$ defined in (4.1.3).

We shall call an estimator *R-optimal* if its expected loss is less than that of any other estimator. More formally, $\hat{\boldsymbol{\beta}}$ is R-optimal if

$$R(\hat{\boldsymbol{\beta}}) \leqq R(\tilde{\boldsymbol{\beta}}) \tag{4.1.5}$$

for any other estimator $\tilde{\boldsymbol{\beta}}$.

Where no ambiguity results we shall use '*optimal*' in place of '*R-optimal*', and of course an '*optimal linear estimator*' will be one defined by (4.1.2) which satisfies (4.1.5) for any other linear estimator $\tilde{\boldsymbol{\beta}}$.

4.2 Optimal estimation

4.2.1 Heterogeneous estimation. From (4.1.1) and (4.1.2) the error in $\hat{\boldsymbol{\beta}}$ is clearly

$$\hat{\boldsymbol{\beta}} - \boldsymbol{\beta} = (\boldsymbol{C}'\boldsymbol{X} - \boldsymbol{I})\,\boldsymbol{\beta} + \boldsymbol{d} + \boldsymbol{C}'\boldsymbol{u}\,. \tag{4.2.1}$$

Therefore

$$R(\hat{\boldsymbol{\beta}}) = [(\boldsymbol{C}'\boldsymbol{X} - \boldsymbol{I})\,\boldsymbol{\beta} + \boldsymbol{d}]'\,\boldsymbol{A}[(\boldsymbol{C}'\boldsymbol{X} - \boldsymbol{I})\,\boldsymbol{\beta} + \boldsymbol{d}] + E[\boldsymbol{u}'\boldsymbol{C}\boldsymbol{A}\boldsymbol{C}'\boldsymbol{u}]\,. \tag{4.2.2}$$

The expression is a multivariate generalization of the well known formula "mean square error equals bias-squared plus variance". The last term in (4.2.2) equals

$$\sigma^2\, tr\, \boldsymbol{A}\boldsymbol{C}'\boldsymbol{W}\boldsymbol{C} \tag{4.2.3}$$

which is free of $\boldsymbol{d}$. Therefore the optimal value of $\boldsymbol{d}$ is that value which minimizes (equates to zero) the first expression in (4.2.2). That occurs when

$$\boldsymbol{d} = \hat{\boldsymbol{d}} = -\,(\hat{\boldsymbol{C}}'\boldsymbol{X} - \boldsymbol{I})\,\boldsymbol{\beta}\,, \tag{4.2.4}$$

where $\hat{\boldsymbol{C}}$ is the yet-to-be-determined optimal value of $\boldsymbol{C}$. Inserting this in (4.2.2) and using (4.2.3), we see that $\hat{\boldsymbol{C}}$ must be the value of $\boldsymbol{C}$ which minimizes $tr\,\boldsymbol{A}\boldsymbol{C}'\boldsymbol{W}\boldsymbol{C}$. Using Theorem A 37 (p. 168),

$$\frac{\partial}{\partial \boldsymbol{C}}\, tr\, \boldsymbol{A}\boldsymbol{C}'\boldsymbol{W}\boldsymbol{C} = 2\boldsymbol{W}\boldsymbol{C}\boldsymbol{A}\,. \tag{4.2.5}$$

This is zero only when $\boldsymbol{C}$ is zero. Substituting this in (4.2.4) yields the trivial conclusion that the *R-optimal heterogeneous estimator is* $\boldsymbol{\beta}$ *itself*. We call this (trivial) estimator $\hat{\boldsymbol{\beta}}_1$. It clearly has zero bias, zero risk — and zero usefulness!

4.2.2 Homogeneous estimation. The R-optimal homogeneous estimator of $\boldsymbol{\beta}$ may be found by putting $\boldsymbol{d} = \boldsymbol{0}$ in (4.1.2) and (4.2.1). This gives

$$\hat{\boldsymbol{\beta}} - \boldsymbol{\beta} = (\boldsymbol{C}'\boldsymbol{X} - \boldsymbol{I})\,\boldsymbol{\beta} + \boldsymbol{C}'\boldsymbol{u} \qquad (4.2.6)$$

Similarly from (4.2.2)

$$R(\hat{\boldsymbol{\beta}}) = \boldsymbol{\beta}'(\boldsymbol{X}'\boldsymbol{C} - \boldsymbol{I})\,\boldsymbol{A}(\boldsymbol{C}'\boldsymbol{X} - \boldsymbol{I})\,\boldsymbol{\beta} + \sigma^2\, tr\, \boldsymbol{A}\boldsymbol{C}'\boldsymbol{W}\boldsymbol{C}\,. \qquad (4.2.7)$$

Differentiating with respect to $\boldsymbol{C}$ gives

$$\frac{\partial R(\boldsymbol{\beta})}{\partial \boldsymbol{C}'} = 2\boldsymbol{A}[\boldsymbol{C}'(\boldsymbol{X}\boldsymbol{\beta}\boldsymbol{\beta}'\boldsymbol{X}' + \sigma^2\boldsymbol{W}) - \boldsymbol{\beta}\boldsymbol{\beta}'\boldsymbol{X}']\,. \qquad (4.2.8)$$

The matrix $(\boldsymbol{X}\boldsymbol{\beta}\boldsymbol{\beta}'\boldsymbol{X}' + \sigma^2\boldsymbol{W})$ is positive definite and invertible. Therefore equating (4.2.8) to zero gives the optimal $\boldsymbol{C}$ as

$$\hat{\boldsymbol{C}}_2' = \boldsymbol{\beta}\boldsymbol{\beta}'\boldsymbol{X}'(\boldsymbol{X}\boldsymbol{\beta}\boldsymbol{\beta}'\boldsymbol{X}' + \sigma^2\boldsymbol{W})^{-1} = \boldsymbol{\beta}\boldsymbol{\beta}'\boldsymbol{X}'\boldsymbol{P}^{-1}\,, \qquad (4.2.9)$$

where $\boldsymbol{P}$ is $(\boldsymbol{X}\boldsymbol{\beta}\boldsymbol{\beta}'\boldsymbol{X}' + \sigma^2\boldsymbol{W})$. Hence we have shown that the *R-optimal homogeneous estimator of* $\boldsymbol{\beta}$ *is*

$$\hat{\boldsymbol{\beta}}_2 = \hat{\boldsymbol{C}}_2'\boldsymbol{y}\,, \qquad (4.2.10)$$

where $\hat{\boldsymbol{C}}_2'$ *is given by* (4.2.9). This expression was derived independently by Toutenburg (1968), Bibby (1972), Rao (1971, p. 389) and Theil (1971, p. 125).

The expression for $\hat{\boldsymbol{C}}_2$ can be considerably simplified by noting that

$$(\sigma^2\boldsymbol{W} + \boldsymbol{X}\boldsymbol{\beta}\boldsymbol{\beta}'\boldsymbol{X}')^{-1} = \sigma^{-2}\boldsymbol{W}^{-1} - \frac{\sigma^{-4}\boldsymbol{W}^{-1}\boldsymbol{X}\boldsymbol{\beta}\boldsymbol{\beta}'\boldsymbol{X}'\boldsymbol{W}^{-1}}{1 + \sigma^{-2}\boldsymbol{\beta}'\boldsymbol{X}'\boldsymbol{W}^{-1}\boldsymbol{X}\boldsymbol{\beta}}\,. \qquad (4.2.11)$$

Equation (4.2.10) then becomes

$$\hat{\boldsymbol{\beta}}_2 = \frac{\sigma^{-2}\boldsymbol{\beta}'\boldsymbol{X}\boldsymbol{W}^{-1}\boldsymbol{y}}{1 + \sigma^{-2}\boldsymbol{\beta}'\boldsymbol{X}'\boldsymbol{W}^{-1}\boldsymbol{X}\boldsymbol{\beta}}\,\boldsymbol{\beta}\,. \qquad (4.2.12)$$

This expression was first given by Bibby (1972) and, independently for the special case $\boldsymbol{W} = \boldsymbol{I}$, by Farebrother (1975).

When $\boldsymbol{\beta}$ is a scalar so that $\boldsymbol{X} = \boldsymbol{x}$ is a vector, then (4.2.12) simplifies to

$$\frac{\boldsymbol{x}'\boldsymbol{y}}{\dfrac{\sigma^2}{\beta^2} + \boldsymbol{x}'\boldsymbol{x}} \qquad (4.2.13)$$

which may be regarded as a 'shrunken' version of the OLSE $\boldsymbol{x}'\boldsymbol{y}/\boldsymbol{x}'\boldsymbol{x}$. Alternatively, putting $\sigma^2\boldsymbol{W} = \boldsymbol{\Sigma}$, (4.2.12) may also be written

$$\hat{\boldsymbol{\beta}}_2 = \frac{\boldsymbol{\beta}'\boldsymbol{X}'\boldsymbol{\Sigma}^{-1}\boldsymbol{y}}{1 + \boldsymbol{\beta}'\boldsymbol{X}'\boldsymbol{\Sigma}^{-1}\boldsymbol{X}\boldsymbol{\beta}}\,\boldsymbol{\beta} \qquad (4.2.14)$$

$$= \boldsymbol{\beta}\left[\boldsymbol{\beta}'\left(\frac{1}{\boldsymbol{\beta}'\boldsymbol{\beta}}\boldsymbol{I} + \boldsymbol{X}'\boldsymbol{\Sigma}^{-1}\boldsymbol{X}\right)\boldsymbol{\beta}\right]^{-1}\boldsymbol{\beta}'\boldsymbol{X}'\boldsymbol{\Sigma}^{-1}\boldsymbol{y}\,. \qquad (4.2.15)$$

Now putting

$$\boldsymbol{V} = \boldsymbol{X}'\boldsymbol{\Sigma}^{-1}\boldsymbol{X} + \frac{1}{\boldsymbol{\beta}'\boldsymbol{\beta}}\boldsymbol{I},$$

(4.2.14) becomes

$$\hat{\boldsymbol{\beta}}_2 = \boldsymbol{\beta}(\boldsymbol{\beta}'\boldsymbol{V}\boldsymbol{\beta})^{-1}\boldsymbol{\beta}'\boldsymbol{V}\boldsymbol{\beta}^+, \tag{4.2.16}$$

where

$$\boldsymbol{\beta}^+ = \left(\boldsymbol{X}'\boldsymbol{\Sigma}^{-1}\boldsymbol{X} + \frac{1}{\boldsymbol{\beta}'\boldsymbol{\beta}}\boldsymbol{I}\right)^{-1}\boldsymbol{X}'\boldsymbol{\Sigma}^{-1}\boldsymbol{y}.$$

FAREBROTHER (1975) gave this expression for the special case $\boldsymbol{\Sigma} = \sigma^2\boldsymbol{I}$. It yields an intuitively pleasing interpretation of $\hat{\boldsymbol{\beta}}_2$ as a two-stage estimator. The first stage calculates $\boldsymbol{\beta}^+$, which is a generalized 'ridge' estimator of the form proposed by HOERL and KENNARD (1970a, 1970b). In the second stage, defined by (4.2.16), $\boldsymbol{\beta}^+$ is regressed on $\boldsymbol{\beta}$, with variance matrix $\boldsymbol{V}^{-1}$. Hence $\hat{\boldsymbol{\beta}}_2$, the homogeneous R-optimal estimator may be interpreted as a combination between the ridge regression procedure and generalized least squares.

Note also from (4.2.14) that

$$E[\hat{\boldsymbol{\beta}}_2] = \boldsymbol{\beta}[1 - (1 + \boldsymbol{\beta}'\boldsymbol{X}'\boldsymbol{\Sigma}^{-1}\boldsymbol{X}\boldsymbol{\beta})^{-1}]. \tag{4.2.18}$$

Hence on average, $\hat{\boldsymbol{\beta}}_2$ leads to an underestimate of $\boldsymbol{\beta}$. Also note that

$$R(\hat{\boldsymbol{\beta}}_1) \leqq R(\hat{\boldsymbol{\beta}}_2). \tag{4.2.19}$$

This inequality may be obtained trivially from the fact that $R(\hat{\boldsymbol{\beta}}_1) = \boldsymbol{0}$ and that $R(\hat{\boldsymbol{\beta}}_2)$ cannot be negative, or from the fact that $\hat{\boldsymbol{\beta}}_2$ is obtained by optimizing over a set of estimators which is a subset of the set that led to $\hat{\boldsymbol{\beta}}_1$. In this example the first derivation of (4.2.19) is the more obvious. However later on in this book the second method, based on constrained optimization, will lead to several useful inequalities. In this way we shall be able to avoid some of the difficulties stemming from the fact that $\hat{\boldsymbol{\beta}}_2$ contains the unknown vector $\boldsymbol{\Sigma}^{-1}\boldsymbol{\beta}$, and is therefore impracticable.

4.2.3 Unbiased homogeneous estimation. From (4.2.1) it is clear that a homogeneous estimator is unbiased if and only if

$$\boldsymbol{C}'\boldsymbol{X} - \boldsymbol{I} = \boldsymbol{0}. \tag{4.2.20}$$

Substituting this in (4.2.7) we find that $R(\hat{\boldsymbol{\beta}})$ becomes $\sigma^2 \, tr\, \boldsymbol{A}\boldsymbol{C}'\boldsymbol{W}\boldsymbol{C}$ when $\hat{\boldsymbol{\beta}}$ is unbiased. Therefore the optimal $\boldsymbol{C}$ in this case is the solution to the following constrained minimization problem:

$$\min_{\boldsymbol{C}} tr\, \boldsymbol{A}\boldsymbol{C}'\boldsymbol{W}\boldsymbol{C} \quad \text{subject to } \boldsymbol{C}'\boldsymbol{X} = \boldsymbol{I}.$$

Adding a Lagrangian matrix Λ and differentiating gives the normal equations

$$\Lambda C'W = \Lambda X'$$

and

$$C'X = I\,.$$

The solution to this is

$$\hat{C}_3' = (X'W^{-1}X)^{-1}\,X'W^{-1} = S^{-1}X'W^{-1}\,, \tag{4.2.21}$$

where $S = X'W^{-1}X$. But $\hat{\beta}_3 = \hat{C}_3'y$ is just the GLS (Gauss-Markov) estimator already defined in (3.4.1). Hence we have shown that the *R-optimal unbiased estimator is also the generalized least squares estimator* $\boldsymbol{b}$.

Using the fact that $\hat{\beta}_3$ is obtained by adding extra conditions to those under which $\hat{\beta}_2$ is optimal, and that $\hat{\beta}_2$ is similarly a more constrained optimum than $\hat{\beta}_1$, it is evident that

$$R(\hat{\beta}_1) \leqq R(\hat{\beta}_2) \leqq R(\hat{\beta}_3) \tag{4.2.22}$$

(see also Section 10.2, especially (10.2.21) to (10.2.22)).

In fact the respective biases and dispersion matrices of these estimators are as follows:

Estimator	Equation	Bias	Dispersion
$\hat{\beta}_1 = \beta$	(4.2.5)	0	0
$\hat{\beta}_2 = \hat{C}_2'y$	(4.2.9), (4.2.10)	$(\hat{C}_2'X - I)\,\beta$	$\sigma^2\beta\beta'X'P^{-1}WP^{-1}X\beta\beta'$
$\hat{\beta}_3 = \hat{C}_3'y = b$	(4.2.21)	0	σ^2S^{-1} .

(4.2.23)

The 'gain' in using the biased estimator $\hat{\beta}_2$, as compared with the unbiased Gauss-Markov estimator $\hat{\beta}_3$ may be evaluated by calculating their respective risks.

Clearly

$$R(\hat{\beta}_3) = \sigma^2\, tr\, AS^{-1} = \sigma^2\, tr\, A(X'W^{-1}X)^{-1}\,.$$

Also

$$R(\hat{\beta}_2) = tr\, A\{\sigma^2\hat{C}_2'W\hat{C}_2 + (\hat{C}_2'X - I)\,\beta\beta'(\hat{C}_2'X - I)'\}\,. \tag{4.2.24}$$

Using (4.2.9) this expression can be simplified, and the relevant comparison can be made.

Finally it is well known that the term σ^2 in (4.1.1) can be consistently estimated by s^2 defined as

$$(n - p)\,s^2 = (y - Xb)'\,W^{-1}(y - Xb)\,. \tag{4.2.25}$$

This estimator is also unbiased. An unbiased and consistent estimator of V_b the covariance matrix of b, is therefore given by

$$\hat{V}_b = s^2S^{-1}\,. \tag{4.2.26}$$

Some work similar to the above is reported by KUPPER and MEYDRECH (1973), who consider estimators of the form $\boldsymbol{K}\hat{\boldsymbol{\beta}}$ where $\boldsymbol{K}$ is a diagonal matrix and $\hat{\boldsymbol{\beta}}$ is the OLS estimator based on a misspecified model. They examine restrictions similar to those which we shall advance in Section 5.4, and also refer to unpublished work by D. C. KORTS which considers a 'maximin' criterion based on 'closeness' concepts similar to Definition 7.2.2.

4.3 Univariate examples

The above results may be illustrated using the univariate model

$$\boldsymbol{y} = \boldsymbol{1}\mu + \boldsymbol{u}\,. \tag{4.3.1}$$

We seek to estimate μ using an estimator of the form

$$\hat{\mu} = \mathbf{c}'\boldsymbol{y} + d\,. \tag{4.3.2}$$

These equations are special cases of (4.1.1) and (4.1.2). The risk function (4.1.3) is simply

$$R(\hat{\mu}) = a^2 E[(\hat{\mu} - \mu)^2] \tag{4.3.3}$$

where a is a constant scalar. In what follows we put $a = 1$. Substituting (4.3.1) in (4.3.2) gives

$$\hat{\mu} = \mathbf{c}'\boldsymbol{1}\mu + \mathbf{c}'\boldsymbol{u} + d\,. \tag{4.3.4}$$

If $\boldsymbol{u}$ has mean zero then

$$E(\hat{\mu}) = \mathbf{c}'\boldsymbol{1}\,\mu + d\,.$$

Therefore the bias is

$$E(\hat{\mu} - \mu) = (\mathbf{c}'\boldsymbol{1} - 1)\,\mu + d\,.$$

For this to be zero

$$d = -\,(\mathbf{c}'\boldsymbol{1} - 1)\,\mu$$

which corresponds to (4.2.4) in the more general case.

The variance of $\hat{\mu}$ from (4.3.4) is

$$\mathbf{c}'\boldsymbol{W}\mathbf{c}$$

where $\boldsymbol{W}$ is the variance of $\boldsymbol{u}$ (taking $\sigma = 1$ in (4.1.1)). Thus the total mean square error of $\hat{\mu}$, equal to (4.3.3) with $a = 1$, is

$$\begin{aligned} MSE\,[\hat{\mu}] &= variance + (bias)^2 \\ &= \mathbf{c}'\boldsymbol{W}\mathbf{c} + [(\mathbf{c}'\boldsymbol{1} - 1)\,\mu + d]^2\,. \end{aligned}$$

This is a special case of (4.2.2), with the order of terms reversed. The value of $\hat{\boldsymbol{c}}_2$ from (4.2.9) is

$$\hat{\boldsymbol{c}}_2' = \mu^2 \boldsymbol{1}'(\mu^2 \boldsymbol{1}\boldsymbol{1}' + \sigma^2 \boldsymbol{W})^{-1}.$$

Using (4.2.11), inserting the values $\boldsymbol{X} = \boldsymbol{1}$ and $\beta = \mu$, we find that

$$\hat{\boldsymbol{c}}_2' = \mu^2 \boldsymbol{1}' \left(\sigma^{-2} \boldsymbol{W}^{-1} - \frac{\mu^2 \sigma^{-4} \boldsymbol{W}^{-1}\boldsymbol{1}\boldsymbol{1}'\boldsymbol{W}^{-1}}{1 + \sigma^{-2}\mu^2 \boldsymbol{1}'\boldsymbol{W}^{-1}\boldsymbol{1}} \right).$$

Note that μ and σ appear in this expression only through the ratio $\sigma/\mu = v$. In terms of v we have

$$\begin{aligned} \hat{\boldsymbol{c}}_2' &= v^{-2} \boldsymbol{1}' \left(\boldsymbol{W}^{-1} - \frac{\boldsymbol{W}^{-1}\boldsymbol{1}\boldsymbol{1}'\boldsymbol{W}^{-1}}{v^2 + \boldsymbol{1}'\boldsymbol{W}^{-1}\boldsymbol{1}} \right) \\ &= \frac{\boldsymbol{1}'\boldsymbol{W}^{-1}}{(v^2 + \boldsymbol{1}'\boldsymbol{W}^{-1}\boldsymbol{1})}. \end{aligned}$$

Thus the estimator $\hat{\boldsymbol{c}}_2'\boldsymbol{y}$ is just a constant times the generalized least squares estimator $\boldsymbol{b}$. When $\boldsymbol{W} = \boldsymbol{I}$ the above expression simplifies to

$$\hat{\boldsymbol{c}}_2' = \frac{1}{v^2 + n}\boldsymbol{1} = \frac{\mu^2}{\sigma^2 + n\mu^2}\boldsymbol{1}$$

which has of course already been met in Chapter 2, and which will also be returned to in Chapter 5.

4.4 Effects of misspecifying the covariance matrix

We now look at the general regression model (4.1.1) and examine the effect of a false choice of covariance matrix $\boldsymbol{W}$ upon the properties of the associated estimators of $\boldsymbol{\beta}$. The reasons for this misspecification could be one of the following:

(a) The correlation structure of disturbances may be unknown, or may have been ignored in order to use OLS estimation and hence simplify calculations.
(b) The correlation structure may be better represented by a matrix which is different from $\boldsymbol{W}$.
(c) The matrix $\boldsymbol{W}$ may be unknown and may have to be estimated.

In any case, our derived estimator will have the form

$$\hat{\boldsymbol{\beta}} = (\boldsymbol{X}'\boldsymbol{A}\boldsymbol{X})^{-1}\boldsymbol{X}'\boldsymbol{A}\boldsymbol{y} \tag{4.4.1}$$

where $\boldsymbol{A} \neq \boldsymbol{W}^{-1}$ (we assume that $\boldsymbol{X}'\boldsymbol{A}\boldsymbol{X}$ is invertible as indicated). It follows that $\hat{\boldsymbol{\beta}}$ is unbiased, and has the covariance matrix

$$\boldsymbol{V}_{\hat{\boldsymbol{\beta}}} = \sigma^2 (\boldsymbol{X}'\boldsymbol{A}\boldsymbol{X})^{-1}\boldsymbol{X}'\boldsymbol{A}\boldsymbol{W}\boldsymbol{A}\boldsymbol{X}(\boldsymbol{X}'\boldsymbol{A}\boldsymbol{X})^{-1}. \tag{4.4.2}$$

Nothing is lost if we suppose that OLS is mistakenly used, and take $\boldsymbol{A}$ to be the identity matrix. Expression (4.4.2) then simplifies to

$$\boldsymbol{V}_{\hat{\beta}} = \sigma^2 \boldsymbol{G}\boldsymbol{X}'\boldsymbol{W}\boldsymbol{X}\boldsymbol{G} \quad \text{where} \quad \boldsymbol{G} = (\boldsymbol{X}'\boldsymbol{X})^{-1} . \tag{4.4.3}$$

The increase in dispersion due to using OLS instead of GLS is

$$\boldsymbol{V}_{\hat{\beta}} - \boldsymbol{V}_{\boldsymbol{b}} = \sigma^2 \boldsymbol{G}\boldsymbol{X}'\boldsymbol{W}\boldsymbol{X}\boldsymbol{G} - \sigma^2 \boldsymbol{S}^{-1} \tag{4.4.4}$$

where $\boldsymbol{S} = \boldsymbol{X}'\boldsymbol{W}^{-1}\boldsymbol{X}$. Following TOUTENBURG (1975, p. 63) this can also be written as

$$\boldsymbol{V}_{\hat{\beta}} - \boldsymbol{V}_{\boldsymbol{b}} = \sigma^2 (\boldsymbol{G}\boldsymbol{X}' - \boldsymbol{S}^{-1}\boldsymbol{X}'\boldsymbol{W}^{-1})\, \boldsymbol{W} (\boldsymbol{G}\boldsymbol{X}' - \boldsymbol{S}^{-1}\boldsymbol{X}'\boldsymbol{W}^{-1})' . \tag{4.4.5}$$

From this formulation it is clear that $(\boldsymbol{V}_{\hat{\beta}} - \boldsymbol{V}_{\boldsymbol{b}})$ is nonnegative definite and moreover that

$$\boldsymbol{V}_{\hat{\beta}} - \boldsymbol{V}_{\boldsymbol{b}} = \boldsymbol{0}$$

when and only when

$$\boldsymbol{G}\boldsymbol{X}' = \boldsymbol{S}^{-1}\boldsymbol{X}'\boldsymbol{W}^{-1}$$

i.e. when

$$\hat{\boldsymbol{\beta}} = \boldsymbol{b} .$$

The conditions for them to be equivalent are given in the following theorem due to McELROY (1967).

Theorem (McELROY) When the regression includes a constant term, the OLSE and GLSE are equivalent when and only when

$$\boldsymbol{W} = \varrho \boldsymbol{I} + (1 - \varrho)\boldsymbol{J} \tag{4.4.6}$$

where $\boldsymbol{J}$ is a matrix of ones and $0 \leqq \varrho \leqq 1$.

In everyday terms, this theorem states that the OLSE and GLSE are identical when, and only when, the variables are equicorrelated and each have the same variance.

For a further analysis of the effects that misspecification can have upon the efficiency of estimators of $\boldsymbol{\beta}$ and σ^2, the reader is referred to GOLDBERGER (1964, p. 238 et seq.), SPRENT (1969, p. 145) and TOUTENBURG (1975, pp. 64–66).

4.5 Heteroscedasticity and autoregression

The disturbances in the general linear model are said to be *heteroscedastic* when they have different variances, and *autoregressive* when they are correlated in a particular manner related to their time sequence. Both these properties negate the assumptions of the GAUSS-MARKOV theorem, and when either property occurs the GLSE cannot in any sense be said to be 'best'.

The usual estimator of σ^2 is (see (4.2.25))

$$s^2 = RSS/(n-p)$$

where RSS is the sum of squared residuals from the GLS regression. When the correct specification is chosen this estimator is unbiased. However problems can arise if an incorrect specification is used. Suppose for instance that OLS is used, when the true covariance matrix is $\sigma^2\boldsymbol{W}$. Then the vector of residuals is

$$\hat{\boldsymbol{u}} = \boldsymbol{y} - \hat{\boldsymbol{y}} = [\boldsymbol{I} - \boldsymbol{X}(\boldsymbol{X}'\boldsymbol{X})^{-1}\boldsymbol{X}']\,\boldsymbol{y} = \boldsymbol{Q}\boldsymbol{y}\,, \quad \text{say}\,, \tag{4.5.1}$$

where $\boldsymbol{Q}$ is the symmetric idempotent matrix in square brackets. Now the residual sum of squares is

$$RSS = \hat{\boldsymbol{u}}'\hat{\boldsymbol{u}} = \boldsymbol{y}'\boldsymbol{Q}\boldsymbol{y}\,.$$

(We use here the fact that $\boldsymbol{Q}$ is idempotent and symmetric, so that $\boldsymbol{Q}'\boldsymbol{Q} = \boldsymbol{Q}$). Now

$$RSS = \boldsymbol{y}'\boldsymbol{Q}\boldsymbol{y} = (\boldsymbol{X}\boldsymbol{\beta} + \boldsymbol{u})'\,\boldsymbol{Q}(\boldsymbol{X}\boldsymbol{\beta} + \boldsymbol{u}) = \boldsymbol{u}'\boldsymbol{Q}\boldsymbol{u} = tr\,\boldsymbol{Q}\boldsymbol{u}\boldsymbol{u}'\,, \tag{4.5.2}$$

since $\boldsymbol{Q}\boldsymbol{X} = \boldsymbol{0}$. Therefore

$$E(\boldsymbol{y}'\boldsymbol{Q}\boldsymbol{y}) = \sigma^2\,tr\,\boldsymbol{Q}\boldsymbol{W} = \sigma^2\,tr\,\boldsymbol{W} - \sigma^2\,tr\,\boldsymbol{X}'\boldsymbol{W}\boldsymbol{X}(\boldsymbol{X}'\boldsymbol{X})^{-1}\,. \tag{4.5.3}$$

This corresponds to equation (9.19) on p. 145 of SPRENT (1969), and in general does *not* equal $(n-p)\,\sigma^2$. Hence in general $s^2 = RSS/(n-p)$ is *not* unbiased. In fact

$$(n-p)\,E[s^2] = \sigma^2\,tr\,\boldsymbol{Q}\boldsymbol{W} = \sigma^2\,tr\,\boldsymbol{Q} + \sigma^2\,tr\,\boldsymbol{Q}(\boldsymbol{W} - \boldsymbol{I})\,.$$

Therefore

$$E[s^2] = \sigma^2 + \frac{\sigma^2}{n-p}\,tr\,\boldsymbol{Q}(\boldsymbol{W} - \boldsymbol{I}) \tag{4.5.4}$$

since $tr\,\boldsymbol{Q} = n - p$. It is usual to standardize $\boldsymbol{W}$ so that its trace is n, in which case

$$tr\,\boldsymbol{Q}(\boldsymbol{W} - \boldsymbol{I}) = tr\,\boldsymbol{X}(\boldsymbol{X}'\boldsymbol{X})^{-1}\,\boldsymbol{X}'(\boldsymbol{I} - \boldsymbol{W}) \tag{4.5.5}$$

by virtue of the fact that $tr\,\boldsymbol{W} = tr\,\boldsymbol{I}$.

The final term in (4.5.4) represents the bias in the estimation of σ^2 when OLS is mistakenly used. A similar bias exists if there is autocorrelation between the disturbance terms.

Note also that a misspecification of $\boldsymbol{W}$ will lead to errors in estimating the covariance matrix of $\hat{\boldsymbol{\beta}}$ (SPRENT 1969, p. 146).

4.6 Summary

This chapter investigates the three standard types of optimal estimator (heterogeneous, homogeneous, and homogeneous unbiased) in the context of the general linear model defined by (4.1.1). The effects of misspecification are examined, in particular the effect of ignoring heteroscedasticity or autoregression in the disturbances.

5

Prediction in linear models

5.1 Classical prediction

We now suppose that in addition to having n observations generated by equation (4.1.1), namely

$$\boldsymbol{y} = \boldsymbol{X\beta} + \boldsymbol{u}\,, \qquad \boldsymbol{u} \sim (\boldsymbol{0}, \sigma^2\boldsymbol{W})\,, \tag{5.1.1}$$

there are also n_* as yet unobserved values of the dependent variable, which may be written as an n_*-vector, $\boldsymbol{y}_*$. This is assumed to be generated by

$$\boldsymbol{y}_* = \boldsymbol{X}_*\boldsymbol{\beta} + \boldsymbol{u}_*\,, \qquad \boldsymbol{u}_* \sim (\boldsymbol{0}, \sigma^2\boldsymbol{W}_*)\,, \tag{5.1.2}$$

where $\boldsymbol{\beta}$ is the same vector as in (5.1.1). In general, $\boldsymbol{X}_*$ is known, and the problem is to estimate $\boldsymbol{y}_*$. We assume that the disturbance terms have means of zero, and covariances given by

$$E[\boldsymbol{uu}'] = \sigma^2\boldsymbol{W}\,, \quad E[\boldsymbol{u}_*\boldsymbol{u}_*'] = \sigma^2\,\boldsymbol{W}_*\,, \quad E[\boldsymbol{uu}_*'] = \sigma^2\boldsymbol{W}_0\,, \tag{5.1.3}$$

where $\boldsymbol{W}$, $\boldsymbol{W}_*$, and $\boldsymbol{W}_0$ are known.

The classical predictor of $\boldsymbol{y}_*$ would be the OLS estimator of $E[\boldsymbol{y}_*]$. This is

$$\hat{\boldsymbol{p}} = \boldsymbol{X}_*\hat{\boldsymbol{\beta}} = \boldsymbol{X}_*\boldsymbol{GX}'\boldsymbol{y}\,, \tag{5.1.4}$$

where $\boldsymbol{G} = (\boldsymbol{X}'\boldsymbol{X})^{-1}$. The expected value of $\hat{\boldsymbol{p}}$ is

$$E[\hat{\boldsymbol{p}}] = \boldsymbol{X}_*\boldsymbol{\beta} = E[\boldsymbol{y}_*]\,.$$

Hence, in the sense that

$$E[\hat{\boldsymbol{p}} - \boldsymbol{y}_*] = \boldsymbol{0}\,, \tag{5.1.5}$$

$\hat{\boldsymbol{p}}$ is unbiased for $\boldsymbol{y}_*$. Note however that the expected value of the predictor $\hat{\boldsymbol{p}}$ cannot *equal* that which is being predicted, $\boldsymbol{y}_*$, since the latter is of course a random variable. The covariance matrix of $\hat{\boldsymbol{p}}$ is

$$\boldsymbol{V}_{\hat{\boldsymbol{p}}} = \sigma^2\boldsymbol{X}_*\boldsymbol{GX}'\boldsymbol{WXGX}_*'\,. \tag{5.1.6}$$

This simplifies when $\boldsymbol{W} = \boldsymbol{I}$ to

$$\sigma^2\boldsymbol{X}_*\boldsymbol{GX}_*'\,. \tag{5.1.7}$$

When $\boldsymbol{W} = \boldsymbol{I}$, the ith element of $\hat{\boldsymbol{p}}$ has variance

$$Var(\hat{p}_i) = \sigma^2 \boldsymbol{x}'_{i*} \boldsymbol{G} \boldsymbol{x}_{i*} \tag{5.1.8}$$

where $\boldsymbol{x}_{i*}$ is the ith row of $\boldsymbol{X}_*$. This variance is largest for a given value of $\boldsymbol{x}'_{i*}\boldsymbol{x}_{i*}$ when $\boldsymbol{x}_{i*}$ is parallel to the eigenvector of $\boldsymbol{G}$ which corresponds to its largest eigenvalue. This is also the eigenvector of $\boldsymbol{X}'\boldsymbol{X}$ which corresponds to its *smallest* eigenvalue. In short, prediction is worst in the directions which have least variation in the elements of $\boldsymbol{X}$. (For the implications of a similar observation in the context of optimal experimental design, see Covey-Crump and Silvey 1970).

This much is standard. However, certain details of the prediction problems might vary.

(a) Attention may centre upon particular elements of $\boldsymbol{y}_*$ — the value at certain fixed points in the future — or we may be interested in the whole vector.
(b) Point predictors may be required, or alternatively interval predictors.
(c) We may possess various types of ancillary information. Alternatively, we may not.

For extensions of this classical approach to prediction the reader is referred to Goldberger (1962) and also the work of Theil (1961, 1966). However our approach to prediction differs from the classical approach in that it develops from the ideas on optimal estimation presented in the previous chapter.

5.2 Optimal prediction

5.2.1 Heterogeneous prediction. It may be argued that the classical approach is not concerned with estimating $\boldsymbol{y}_*$ so much as its expected value, $\boldsymbol{X}_*\boldsymbol{\beta}$. In this sense the 'unbiasedness' defined by (5.1.5) was somewhat unsatisfactory. Therefore we now turn to examine the prediction of $\boldsymbol{y}_*$ itself, looking especially at methods of using the correlation structure specified in (5.1.3) to improve the classical predictor.

In a similar manner to Chapter 4 we consider first a linear heterogeneous predictor having the form

$$\boldsymbol{p} = \boldsymbol{C}'\boldsymbol{y} + \boldsymbol{d}\,, \tag{5.2.1}$$

and a quadratic risk function

$$R(\boldsymbol{p}) = E[(\boldsymbol{p} - \boldsymbol{y}_*)'\,\boldsymbol{A}(\boldsymbol{p} - \boldsymbol{y}_*)]\,. \tag{5.2.2}$$

In this equation the matrix $\boldsymbol{A}$ is non-negative definite, and may be chosen to reflect individual circumstances. For instance tomorrow may be more relevant than next year — earlier predictions may be more important than later ones. If this is so, we could take

$$\boldsymbol{A} = diag\,(a_1, a_2, \dots, a_n) \tag{5.2.3}$$

where

$$a_1 > a_2 > \cdots > a_n > 0 .$$

In this special case the loss function given by (5.2.2) has the particularly simple form

$$R(p) = \sum_i a_i E[(p_i - y_{i*})^2] .$$

Returning now to the general problem, we seek to minimize (5.2.2) with respect to $\boldsymbol{C}$ and $\boldsymbol{d}$, which define $\boldsymbol{p}$ in (5.2.1). Using (5.1.2) we have

$$\left.\begin{aligned} \boldsymbol{p} - \boldsymbol{y}_* &= \boldsymbol{C}'\boldsymbol{y} + \boldsymbol{d} - \boldsymbol{X}_*\boldsymbol{\beta} - \boldsymbol{u}_* \\ &= (\boldsymbol{C}'\boldsymbol{X} - \boldsymbol{X}_*)\,\boldsymbol{\beta} + \boldsymbol{d} + (\boldsymbol{C}'\boldsymbol{u} - \boldsymbol{u}_*) . \end{aligned}\right\} \tag{5.2.4}$$

The risk function (5.2.2) is therefore

$$\begin{aligned} \tilde{R}(\boldsymbol{C}, \boldsymbol{d}) &= tr\,\boldsymbol{A}[(\boldsymbol{C}'\boldsymbol{X} - \boldsymbol{X}_*)\,\boldsymbol{\beta} + \boldsymbol{d}]\,[(\boldsymbol{C}'\boldsymbol{X} - \boldsymbol{X}_*)\,\boldsymbol{\beta} + \boldsymbol{d}]' \\ &\quad + \sigma^2\, tr\,\boldsymbol{A}[\boldsymbol{C}'\boldsymbol{W}\boldsymbol{C} + \boldsymbol{W}_* - 2\boldsymbol{C}'\boldsymbol{W}_0] . \end{aligned} \tag{5.2.5}$$

Of these two terms, only the first involves $\boldsymbol{d}$. (Note the similarity with (4.2.2).) Therefore the risk function is minimized when

$$\boldsymbol{d} = (\boldsymbol{X}_* - \boldsymbol{C}'\boldsymbol{X})\,\boldsymbol{\beta} . \tag{5.2.6}$$

To find the optimal value of $\boldsymbol{C}$ the final term in (5.2.5) is now differentiated with respect to $\boldsymbol{C}$. This gives

$$\boldsymbol{A}\boldsymbol{C}'\boldsymbol{W} = \boldsymbol{A}\boldsymbol{W}_0' .$$

Therefore

$$\boldsymbol{C} = \boldsymbol{W}^{-1}\boldsymbol{W}_0 = \hat{\boldsymbol{C}}_1 , \quad \text{say} . \tag{5.2.7}$$

(We assume here that $\boldsymbol{A}$ and $\boldsymbol{W}$ are positive definite and therefore invertible.) Substituting (5.2.7) into (5.2.6) gives

$$\boldsymbol{d} = (\boldsymbol{X}_* - \boldsymbol{W}_0'\boldsymbol{W}^{-1}\boldsymbol{X})\,\boldsymbol{\beta} = \hat{\boldsymbol{d}}_1 , \quad \text{say} . \tag{5.2.8}$$

Thus we have proved the following.

Theorem 5.1 In the model specified by (5.1.1) to (5.1.3), the R-optimal heterogeneous predictor of $\boldsymbol{y}_*$ is

$$\hat{\boldsymbol{p}}_1 = \hat{\boldsymbol{C}}_1'\boldsymbol{y} + \hat{\boldsymbol{d}}_1 = \boldsymbol{X}_*\boldsymbol{\beta} + \boldsymbol{W}_0'\boldsymbol{W}^{-1}(\boldsymbol{y} - \boldsymbol{X}\boldsymbol{\beta}) . \tag{5.2.9}$$

Note that $\boldsymbol{p}_1$ has expectation $\boldsymbol{X}_*\boldsymbol{\beta}$ and therefore is unbiased in the sense defined by (5.1.5). The risk of $\hat{\boldsymbol{p}}_1$ is

$$R(\hat{\boldsymbol{p}}_1) = \sigma^2\, tr\,\boldsymbol{A}(\boldsymbol{W}_* - \boldsymbol{W}_0'\boldsymbol{W}^{-1}\boldsymbol{W}_0) . \tag{5.2.10}$$

Also of course

$$R(\hat{\boldsymbol{p}}_1) \leqq R(\hat{\boldsymbol{p}}),$$

where $\hat{\boldsymbol{p}}$ is the classical predictor given by (5.1.4). However these results are of little practical use, since $\hat{\boldsymbol{p}}_1$ depends upon the unknown vector $\boldsymbol{\beta}$. Nevertheless, (5.2.10) provides a lower bound against which to compare other practicable estimators.

Various approaches to practicability are possible. We could estimate $\boldsymbol{\beta}$ initially by any arbitrary estimator $\boldsymbol{b}_0$ using prior information, generalized least squares, or some other estimation procedure, inserting this initial estimator into (5.2.9). This would give the predictor

$$\hat{\boldsymbol{p}}_1(\boldsymbol{b}_0) = \boldsymbol{X}_*\boldsymbol{b}_0 + \boldsymbol{W}_0'\boldsymbol{W}^{-1}(\boldsymbol{y} - \boldsymbol{X}\boldsymbol{b}_0). \tag{5.2.11}$$

In general we would expect that

$$R[\hat{\boldsymbol{p}}_1(\boldsymbol{b}_0)] \geqq R(\hat{\boldsymbol{p}}_1), \tag{5.2.12}$$

since (5.2.10) gives a lower bound on $R[\hat{\boldsymbol{p}}_1(\boldsymbol{b}_0)]$.

Note that if the OLSE $\hat{\boldsymbol{\beta}}$ is used for $\boldsymbol{b}_0$ in (5.2.11), then

$$\hat{\boldsymbol{p}}_1(\hat{\boldsymbol{\beta}}) = \boldsymbol{X}_*\hat{\boldsymbol{\beta}} + \boldsymbol{W}_0'\boldsymbol{W}^{-1}(\boldsymbol{y} - \boldsymbol{X}\hat{\boldsymbol{\beta}}). \tag{5.2.13}$$

This may be interpreted as the classical predictor (5.1.4), adjusted by a term depending on the residuals in the already observed time periods, $(\boldsymbol{y} - \boldsymbol{X}\hat{\boldsymbol{\beta}})$, and the assumed covariance between the disturbances which these approximate and the later disturbances $(\sigma^2\boldsymbol{W}_0)$. Note also that (5.2.13) is a particular case of the '*best linear unbiased predictor*' proposed by Goldberger (1962, equation 3.13).

5.2.2 Homogeneous prediction. We now turn to the special case of (5.2.1) where $\boldsymbol{d}$ is zero. That is we consider predictors of the form

$$\boldsymbol{p} = \boldsymbol{C}'\boldsymbol{y}. \tag{5.2.14}$$

The risk function (5.2.2) then becomes

$$\tilde{R}(\boldsymbol{C}) = \boldsymbol{\beta}'(\boldsymbol{C}'\boldsymbol{X} - \boldsymbol{X}_*)'\boldsymbol{A}(\boldsymbol{C}'\boldsymbol{X} - \boldsymbol{X}_*)\boldsymbol{\beta} + \sigma^2\, tr\, \boldsymbol{A}[\boldsymbol{C}'\boldsymbol{W}\boldsymbol{C} + \boldsymbol{W}_* - 2\boldsymbol{C}'\boldsymbol{W}_0]. \tag{5.2.15}$$

The optimal value of $\boldsymbol{C}$ is found by differentiating and equating to zero. This leads to the optimal homogeneous predictor

$$\hat{\boldsymbol{p}}_2 = \hat{\boldsymbol{C}}_2'\boldsymbol{y} \tag{5.2.16}$$

where $\hat{\boldsymbol{C}}_2 = (\sigma^{-2}\boldsymbol{X}\boldsymbol{\beta}\boldsymbol{\beta}'\boldsymbol{X}' + \boldsymbol{W})^{-1}(\boldsymbol{W}_0 + \sigma^{-2}\boldsymbol{X}\boldsymbol{\beta}\boldsymbol{\beta}'\boldsymbol{X}_*')$. (For details see Toutenburg 1975, p. 71).

5.2.3 Unbiased homogeneous prediction. Just as in the case of the heterogeneous predictor, the optimal homogeneous predictor defined by (5.2.16) is not practicable since it contains the unknown vector $\sigma^{-1}\boldsymbol{\beta}$. However this difficulty can be removed if we add to (5.2.14) the condition that $\boldsymbol{p}$ should be unbiased, since then

$$\boldsymbol{0} = E[\boldsymbol{p} - \boldsymbol{y}_*] = (\boldsymbol{C}'\boldsymbol{X} - \boldsymbol{X}_*)\,\boldsymbol{\beta}\,.$$

This condition is satisfied if and only if

$$\boldsymbol{C}'\boldsymbol{X} = \boldsymbol{X}_*\,. \tag{5.2.17}$$

Adding this to (5.2.15) we find that the risk function for an unbiased homogeneous predictor is

$$\tilde{R}(\boldsymbol{C}) = \sigma^2\, tr\, \boldsymbol{A}(\boldsymbol{C}'\boldsymbol{W}\boldsymbol{C} + \boldsymbol{W}_* - 2\boldsymbol{C}'\boldsymbol{W}_0)\,. \tag{5.2.18}$$

This is minimized by inserting a set of Lagrangian multipliers $\boldsymbol{\Lambda}$ and considering the following problem:

$$\min_{\boldsymbol{C}}\, \{\tilde{R}(\boldsymbol{C}) - 2\boldsymbol{\Lambda}(\boldsymbol{C}'\boldsymbol{X} - \boldsymbol{X}_*)\}\,. \tag{5.2.19}$$

The normal equations obtained by differentiating are

$$\boldsymbol{0} = \boldsymbol{A}\boldsymbol{C}'\boldsymbol{W} - \boldsymbol{A}\boldsymbol{W}_0' - \sigma^2\boldsymbol{\Lambda}\boldsymbol{X}' \tag{5.2.20}$$

and

$$\boldsymbol{0} = \boldsymbol{C}'\boldsymbol{X} - \boldsymbol{X}_*\,.$$

(For details see Toutenburg 1975, p. 72). From (5.2.20) it follows that the optimal unbiased homogeneous predictor is

$$\hat{\boldsymbol{p}}_3 = \hat{\boldsymbol{C}}_3'\boldsymbol{y} \tag{5.2.21}$$

where

$$\hat{\boldsymbol{C}}_3' = \boldsymbol{X}_*'\boldsymbol{S}^{-1}\boldsymbol{X}'\boldsymbol{W}^{-1} + \boldsymbol{W}_0'\boldsymbol{W}^{-1}(\boldsymbol{I} - \boldsymbol{X}\boldsymbol{S}^{-1}\boldsymbol{X}'\boldsymbol{W}^{-1})\,.$$

Note that $\hat{\boldsymbol{C}}_3'\boldsymbol{X} = \boldsymbol{X}_*$ and therefore $\hat{\boldsymbol{p}}_3$ can also be written

$$\hat{\boldsymbol{p}}_3 = \boldsymbol{X}_*\boldsymbol{b} + \boldsymbol{W}_0'\boldsymbol{W}^{-1}(\boldsymbol{y} - \boldsymbol{X}\boldsymbol{b}) \tag{5.2.22}$$

where $\boldsymbol{b}$ is the GLSE. This expression is just $\hat{\boldsymbol{p}}_1(\boldsymbol{b})$, where $\boldsymbol{p}_1(.)$ is given by (5.2.13).

In practice, when there is no auxiliary information available, $\hat{\boldsymbol{p}}_3$ may be regarded as the most sensible predictor to use.

Summarizing the above results, we have the following (see 4.2.25 and 5.2.9).

Theorem 5.2 With the specification defined in (5.1.1) to (5.1.3) the R-optimal homogeneous predictors of $\boldsymbol{y}_*$ are $\hat{\boldsymbol{p}}_2$ given by (5.2.16) for the biased predictor, and $\hat{\boldsymbol{p}}_3$ given by (5.2.21) for the unbiased predictor. The corresponding values of the risk function are

$$R(\hat{\boldsymbol{p}}_2) = \tilde{R}(\hat{\boldsymbol{C}}_2)$$

with $\tilde{R}(.)$ given by (5.2.18) and

$$R(\hat{\boldsymbol{p}}_3) = R(\hat{\boldsymbol{p}}_1) + \sigma^2\, tr\, \boldsymbol{A}(\boldsymbol{X}_* - \boldsymbol{W}_0'\boldsymbol{W}^{-1}\boldsymbol{X})\,\boldsymbol{S}^{-1}(\boldsymbol{X}_* - \boldsymbol{W}_0'\boldsymbol{W}^{-1}\boldsymbol{X})' \tag{5.2.23}$$

where $R(\hat{\boldsymbol{p}}_1)$ is defined by (5.2.10).

Note that neither $\hat{\boldsymbol{p}}_2$ nor $\hat{\boldsymbol{p}}_3$ involves the matrix $\boldsymbol{A}$. Also the risks of these predictors satisfy the following relationship:

$$R(\hat{\boldsymbol{p}}_1) \leqq R(\hat{\boldsymbol{p}}_2) \leqq R(\hat{\boldsymbol{p}}_3) . \qquad (5.2.24)$$

5.3 Some general comments

5.3.1 Relationships between estimation and prediction. We now turn to compare these results on optimal prediction with those concerning optimal estimation derived in Chapter 4. Similarities clearly exist between the risk functions specified by (4.1.3) and (5.2.2). This suggests a close mathematical relationship between optimal prediction and estimation (see TOUTENBURG 1968). This relationship is formalized in the following theorem.

Theorem 5.3 In the general linear regression model the R-optimal estimators give the optimal classical predictors $\check{\boldsymbol{p}}_1 = \boldsymbol{X}_* \hat{\boldsymbol{\beta}}_1$ (with $\hat{\boldsymbol{\beta}}_1 = \boldsymbol{\beta}$ the optimal heterogeneous estimator), $\check{\boldsymbol{p}}_2 = \boldsymbol{X}_* \hat{\boldsymbol{\beta}}_2 = \boldsymbol{X}_* \hat{\boldsymbol{C}}_2' \boldsymbol{y}$ (see (4.2.10)), $\check{\boldsymbol{p}}_3 = \boldsymbol{X}_* \hat{\boldsymbol{\beta}}_3$ (here $\hat{\boldsymbol{\beta}}_3 = \boldsymbol{b}$, the GLS estimator). These predictors have the following properties:

$$R(\check{\boldsymbol{p}}_1) = \sigma^2 \, tr \, \boldsymbol{A}\boldsymbol{W}_* ; \qquad (5.3.1)$$

$$\begin{aligned} R(\check{\boldsymbol{p}}_2) = R(\check{\boldsymbol{p}}_1) &+ tr \, \boldsymbol{A}\boldsymbol{X}_* \boldsymbol{V}_{\hat{\boldsymbol{\beta}}_2} \boldsymbol{X}_*' + \\ &+ tr \, \boldsymbol{A}\{\boldsymbol{X}_* [bias \, \hat{\boldsymbol{\beta}}_2] \, [bias \, \hat{\boldsymbol{\beta}}_2]' \, \boldsymbol{X}_*' - 2\sigma^2 \boldsymbol{X}_* \hat{\boldsymbol{C}}_2' \boldsymbol{W}_0\} , \end{aligned} \qquad (5.3.2)$$

$$R(\check{\boldsymbol{p}}_3) = R(\check{\boldsymbol{p}}_1) + tr \, \boldsymbol{A}\boldsymbol{X}_* \boldsymbol{V}_{\hat{\boldsymbol{\beta}}_3} \boldsymbol{X}_*' - 2\sigma^2 \, tr \, \boldsymbol{A}\boldsymbol{X}_* \boldsymbol{S}^{-1} \boldsymbol{X}' \boldsymbol{W}^{-1} \boldsymbol{W}_0 . \qquad (5.3.3)$$

The values of $\hat{\boldsymbol{\beta}}_1$, $\hat{\boldsymbol{\beta}}_2$, and $\hat{\boldsymbol{\beta}}_3$ are given in (4.2.23). (For a proof of this theorem see TOUTENBURG (1975, p. 74)).

5.3.2 Gains in efficiency. Note that when $\boldsymbol{W}_0 = \boldsymbol{0}$ the R-optimal predictors are equivalent to the classical predictors. However when $\boldsymbol{W}_0 \neq \boldsymbol{0}$ we may investigate how much efficiency is gained by using the information contained in $\boldsymbol{W}_0$ for estimation purposes. The gain in efficiency is evaluated by comparing the risk of the R-optimal predictor with that of the classical predictor.

(a) Heterogeneous predictor: From (5.2.10) and (5.3.1) we obtain

$$R(\check{\boldsymbol{p}}_1) - R(\hat{\boldsymbol{p}}_1) = \sigma^2 \, tr \boldsymbol{A}\boldsymbol{W}_0' \boldsymbol{W}^{-1} \boldsymbol{W}_0 = E[\boldsymbol{u}' \boldsymbol{W}^{-1} \boldsymbol{W}_0 \boldsymbol{A} \boldsymbol{W}_0' \boldsymbol{W}^{-1} \boldsymbol{u}] \geqq 0 \qquad (5.3.4)$$

since this quadratic form is non-negative definite. Therefore this difference has a non-negative expectation.

(b) Homogeneous predictor: (5.2.15), (5.2.16), and (5.3.2) yield the relationship

$$R(\check{\boldsymbol{p}}_2) - R(\hat{\boldsymbol{p}}_2) = \sigma^2 \, tr \, \boldsymbol{A}\boldsymbol{W}_0' (\sigma^{-2} \boldsymbol{X}\boldsymbol{\beta}\boldsymbol{\beta}' \boldsymbol{X}' + \boldsymbol{W})^{-1} \boldsymbol{W}_0 \geqq 0 .$$

(c) Unbiased homogeneous predictor: From (5.2.18), (5.2.21), and (5.3.3)

$$R(\check{\boldsymbol{p}}_3) - R(\hat{\boldsymbol{p}}_3) = \sigma^2 \, tr \, \boldsymbol{A}\boldsymbol{W}_0'(\boldsymbol{W}^{-1} - \boldsymbol{W}^{-1}\boldsymbol{X}\boldsymbol{S}^{-1}\boldsymbol{X}'\boldsymbol{W}^{-1}) \, \boldsymbol{W}_0$$
$$= E\tilde{\boldsymbol{u}}'\boldsymbol{A}\tilde{\boldsymbol{u}} \geqq 0$$

where

$$\tilde{\boldsymbol{u}}' = \boldsymbol{v}'(\boldsymbol{N}' - \boldsymbol{N}'\boldsymbol{X}'\boldsymbol{S}'^{-1}\boldsymbol{X}'\boldsymbol{W}^{-1}) \, \boldsymbol{W}_0 \,, \qquad \boldsymbol{N}\boldsymbol{N}' = \boldsymbol{W}^{-1}$$

and $\boldsymbol{v}$ is a random variable with $E[\boldsymbol{v}\boldsymbol{v}'] = \sigma^2\boldsymbol{I}$.

5.4 Auxiliary information in the form of inequalities

5.4.1 General results. As we have seen, the optimal homogeneous predictor given by (5.2.16) contains the unknown parameter vector $\sigma^{-1}\boldsymbol{\beta}$ and is therefore not practicable without further assumptions or auxiliary information. This section examines how the assumption of inequalities on $\sigma^{-1}\boldsymbol{\beta}$ can be utilized in the context of prediction. The work of Läuter (1970) has indicated that if the only restriction concerns the direction of $\boldsymbol{X\beta}$, then for any homogeneous predictor $p = \boldsymbol{c}'\boldsymbol{y}$ there is no heterogeneous predictor $p + d$ which is uniformly better than p. In other words, no d exists such that $R(p + d) < R(p)$. Therefore, in order to guarantee an improvement, the prior information must provide more information than that contained in the direction of $\boldsymbol{X\beta}$.

Toutenburg (1970b) proposed a solution which will be examined below. We concentrate upon the problem of predicting a single observation i.e. $n_* = 1$. The risk function in this case is then simply

$$R(p) = E(p - y_*)^2$$

where

$$y_* = \boldsymbol{x}_*'\boldsymbol{\beta} + u_*$$

and

$$E[u_*] = 0 \,, \qquad E[u_*^2] = \sigma_*^2 \,, \qquad E[\boldsymbol{u}u_*] = \sigma^2\boldsymbol{w} \,. \tag{5.4.1}$$

(These equations are special cases of (5.1.2) and (5.1.3) with $\boldsymbol{x}_*$ a $p \times 1$-vector, $\boldsymbol{w}$ a $p \times 1$-vector and y_*, u_*, σ_*^2 scalars.)

Suppose that we know a fixed vector $\boldsymbol{\beta}_0$ such that

$$\sigma^{-1}(\boldsymbol{\beta} - \boldsymbol{\beta}_0) = \sigma^{-1}\boldsymbol{\delta}$$

is bounded in both direction and length. We take a special heterogeneous set-up for the predictor as

$$p_1^\delta = \boldsymbol{c}'\boldsymbol{y} - (\boldsymbol{c}'\boldsymbol{X} - \boldsymbol{x}_*') \, \boldsymbol{\beta}_0 \,. \tag{5.4.2}$$

The risk of this predictor depends directly on $\boldsymbol{\delta}$ as follows:

$$R(p_1^\delta) = [(\boldsymbol{c}'\boldsymbol{X} - \boldsymbol{x}_*') \, \boldsymbol{\delta}]^2 + \sigma^2\boldsymbol{c}'\boldsymbol{W}\boldsymbol{c} + \sigma_*^2 - 2\sigma^2\boldsymbol{c}'\boldsymbol{w} \,. \tag{5.4.3}$$

Using (5.2.16) we get as our optimal predictor

$$\hat{p}_1^\delta = \boldsymbol{c}'^\delta(\boldsymbol{y} - \boldsymbol{X}\boldsymbol{\beta}_0) + \boldsymbol{x}'_*\boldsymbol{\beta}_0 \tag{5.4.4}$$

where $\boldsymbol{c}'^\delta$ equals

$$(\boldsymbol{w}' + \sigma^{-2}\boldsymbol{x}'_*\boldsymbol{\delta}\boldsymbol{\delta}'\boldsymbol{X}')\,(\sigma^{-2}\boldsymbol{X}\boldsymbol{\delta}\boldsymbol{\delta}'\boldsymbol{X}' + \boldsymbol{W})^{-1}\,. \tag{5.4.5}$$

This predictor has a risk function which equals (5.4.3), with $\boldsymbol{c}^\delta$ substituted for $\boldsymbol{c}$. Note that $\hat{p}_1^\delta$ depends on the values of $\boldsymbol{\beta}_0$ and $\boldsymbol{\delta}$. However, the following result enables one to derive a practicable improved predictor which is better than $\hat{p}_2$. This theorem depends upon the knowledge of a vector $\boldsymbol{\mu}$ which exceeds the value of $\sigma^{-1}\boldsymbol{\delta}$ in each of its elements.

Theorem 5.4 Let there be known vectors $\boldsymbol{\mu}$ and $\boldsymbol{\beta}_0$ such that

$$\sigma^{-1}(\boldsymbol{\beta} - \boldsymbol{\beta}_0) = \sigma^{-1}\boldsymbol{\delta} \leqq \boldsymbol{\mu}\,. \tag{5.4.6}$$

Further, let there exist a known vector $\boldsymbol{\Delta}$ such that

$$\sum_i |\{\hat{\boldsymbol{c}}'^\Delta\boldsymbol{X} - \boldsymbol{x}'_*\}_i|\,\mu_i \leqq |\sum_i \{\hat{\boldsymbol{c}}'^\Delta\boldsymbol{X} - \boldsymbol{x}'_*\}_i\,\Delta_i| \tag{5.4.7}$$

and

$$\sigma^{-1}\boldsymbol{\beta} = \alpha\boldsymbol{\Delta} \qquad \text{where } \alpha^2 > 1\,. \tag{5.4.8}$$

Then the predictor

$$\hat{p}_1^\Delta = \hat{\boldsymbol{c}}_1'^\Delta(\boldsymbol{y} - \boldsymbol{X}\boldsymbol{\beta}_0) + \boldsymbol{x}'_*\boldsymbol{\beta}_0 \tag{5.4.9}$$

where

$$\hat{\boldsymbol{c}}'^\Delta = (\boldsymbol{w}' + \boldsymbol{x}'_*\boldsymbol{\Delta}\boldsymbol{\Delta}'\boldsymbol{X}')\,(\boldsymbol{X}\boldsymbol{\Delta}\boldsymbol{\Delta}'\boldsymbol{X}' + \boldsymbol{W})^{-1} \tag{5.4.10}$$

is practicable and more efficient than $\hat{p}_2$ i.e.

$$R(\hat{p}_1^\Delta) < R(\hat{p}_2)\,. \tag{5.4.11}$$

The proof of this theorem is complicated and therefore omitted. It is given in Toutenburg (1975, pp. 76—79). Its primary implication is that when $\boldsymbol{\beta}$ is bounded in both direction and size, then the classical predictor can be improved. Note, however, that when $\boldsymbol{\beta}_0 = \boldsymbol{0}$, the inequality in (5.4.6) relates just to the size of $\boldsymbol{\beta}$, not to its direction. In these circumstances $R(\hat{p}_1^\delta)$ equals $R(\hat{p}_2)$, so that no improvement is possible using this method.

At the other extreme, if $\boldsymbol{\beta} = \boldsymbol{\beta}_0$ is known exactly then $\boldsymbol{\mu}$ in (5.4.6) is the null vector, $\boldsymbol{\Delta}$ in (5.4.7) is zero, and

$$\hat{p}_1^\Delta = \hat{p}_1^0 = \hat{p}_1\,.$$

5.4.2 The Cobb-Douglas function: an economic example. Note that (5.4.8) is a sufficient condition for the validity of the theorem, but is not in fact necessary. Nevertheless, (5.4.8) is satisfied in several practical situations. For instance, consider the Cobb-Douglas production function with a multiplicative disturbance term. This has the form

$$Y = A^{\beta_1}B^{\beta_2}e^u \tag{5.4.12}$$

where A and B measure inputs (capital, labour, etc.), and Y relates to the output of a particular production process. The coefficients β_1 and β_2 are the so-called coefficients of elasticity, which relate the proportional change in output to the proportional change in input. For instance

$$\beta_1 = \frac{A}{Y} \frac{\partial Y}{\partial A} .$$

Now the ratio of elasticity coefficients, $\beta_1/\beta_2 = k$ is often known. Furthermore if we assume that the production process is subject to increasing returns to scale then $\beta_1 + \beta_2 > 1$. If we now choose a vector $\boldsymbol{\Delta}$ such that

$$\boldsymbol{\Delta} = (\Delta_1, \Delta_2)' = (\Delta_1, k\Delta_1)'$$

where

$$\Delta_1(1 + k) < 1 ,$$

then (5.4.1) and (5.4.8) are satisfied and the improved estimator given by (5.4.9) can be used.

5.4.3 The case of a single regressor variable. If each value of the dependent variable depends upon just one regressor, then the results of the previous section simplify. For instance, then

$$\hat{p}_2 = \hat{\boldsymbol{c}}_2'\boldsymbol{y}$$

with

$$\hat{\boldsymbol{c}}_2' = \boldsymbol{w}'\boldsymbol{W}^{-1} + \frac{\sigma^{-2}\beta^2(\boldsymbol{x}_* - \boldsymbol{w}'\boldsymbol{W}^{-1}\boldsymbol{X})\,\boldsymbol{X}'\boldsymbol{W}^{-1}}{(1 + \sigma^{-2}\beta^2\boldsymbol{X}'\boldsymbol{W}^{-1}\boldsymbol{X})} , \tag{5.4.13}$$

and

$$R(\hat{p}_2) = \sigma_*^2 - \sigma^2\boldsymbol{w}'\boldsymbol{W}^{-1}\boldsymbol{w} + \frac{\beta^2(\boldsymbol{x}_* - \boldsymbol{w}'\boldsymbol{W}^{-1}\boldsymbol{X})^2}{(1 + \sigma^{-2}\beta^2\boldsymbol{X}'\boldsymbol{W}^{-1}\boldsymbol{X})^2} .$$

Also

$$\hat{p}_1 = \hat{\boldsymbol{c}}_1'\boldsymbol{y} + d = \hat{\boldsymbol{c}}_1'(\boldsymbol{y} - \boldsymbol{X\beta}) + \boldsymbol{x}_*\boldsymbol{\beta}$$

with

$$\hat{\boldsymbol{c}}_1' = \boldsymbol{w}'\boldsymbol{W}^{-1}$$

and

$$R(\hat{p}_1) = \sigma_*^2 - \sigma^2\boldsymbol{w}'\boldsymbol{W}^{-1}\boldsymbol{w} ,$$

and

$$\hat{p}_1^\delta = \hat{\boldsymbol{c}}'^\delta(\boldsymbol{y} - \boldsymbol{X\beta}_0) + x_*'\boldsymbol{\beta}_0 \tag{5.4.14}$$

with

$$\hat{\boldsymbol{c}}'^\delta = \boldsymbol{w}'\boldsymbol{W}^{-1} + \frac{\sigma^{-2}\delta^2(x_* - \boldsymbol{w}'\boldsymbol{W}^{-1}\boldsymbol{X})\,\boldsymbol{X}'\boldsymbol{W}^{-1}}{(1 + \sigma^{-2}\delta^2\boldsymbol{X}'\boldsymbol{W}^{-1}\boldsymbol{X})} ,$$

and

$$R(\hat{p}_1^\delta) = \sigma_*^2 - \sigma^2\boldsymbol{w}'\boldsymbol{W}^{-1}\boldsymbol{w} + \frac{\delta^2(x_* - \boldsymbol{w}'\boldsymbol{W}^{-1}\boldsymbol{X})^2}{(1 + \sigma^{-2}\delta^2\boldsymbol{X}'\boldsymbol{W}^{-1}\boldsymbol{X})^2} .$$

We assume that we know as auxiliary information the fact that there exists a positive scalar Δ such that

$$|\sigma^{-1}\delta| \leqq \Delta < |\sigma^{-1}\beta| \,, \tag{5.4.15}$$

where $\delta = \beta - \beta_0$ and $\beta_0 \neq 0$. The two inequalities in (5.4.15) are compatible if for instance

$$\sigma^{-1}\,|\beta_0| > 2\Delta \,.$$

Substituting Δ for δ in (5.4.14) gives the predictor

$$\hat{p}_1^\Delta = \hat{\boldsymbol{c}}'^\Delta(\boldsymbol{y} - \boldsymbol{X}\beta_0) + x_*\beta_0 \tag{5.4.16}$$

with

$$\hat{\boldsymbol{c}}'^\Delta = \boldsymbol{w}'\boldsymbol{W}^{-1} + \frac{\Delta^2(x_* - \boldsymbol{w}'\boldsymbol{W}^{-1}\boldsymbol{X})\,\boldsymbol{X}'\boldsymbol{W}^{-1}}{(1 + \Delta^2\boldsymbol{X}'\boldsymbol{W}^{-1}\boldsymbol{X})} \,.$$

Writing $f(\sigma^{-1}\beta)$ for $R(\hat{p}_2)$ we note from (5.4.15) that

$$R(\hat{p}_1^\delta) \leqq R(\hat{p}_1^\Delta) \leqq f(\Delta) < f(\sigma^{-1}\beta) = R(\hat{p}_2) \,. \tag{5.4.17}$$

In other words, the inequalities given by (5.4.15) have led to a predictor $\hat{p}_1^\Delta$ which is distinctly better than $\hat{p}_2$ and yet which contains no unknown parameters. When Δ is zero, so that $\beta = \beta_0$ is exactly known, then $\hat{p}_1^\Delta$ has the same risk as $\hat{p}_1$. If only the left hand side inequality of (5.4.15) holds we can no longer say that $\hat{p}_1^\Delta$ is better than $\hat{p}_2$, although it is certainly still better than $\hat{p}_3$. This may be shown as follows. If

$$\hat{p}_3 = \boldsymbol{c}_3'\boldsymbol{y}$$

where

$$\hat{\boldsymbol{c}}_3' = \boldsymbol{w}'\boldsymbol{W}^{-1} + \frac{(x_* - \boldsymbol{w}'\boldsymbol{W}^{-1}\boldsymbol{X})\,\boldsymbol{X}'\boldsymbol{W}^{-1}}{\boldsymbol{X}'\boldsymbol{W}^{-1}\boldsymbol{X}} \,,$$

then

$$R(\hat{p}_3) = \sigma_*^2 - \sigma^2\boldsymbol{w}'\boldsymbol{W}^{-1}\boldsymbol{w} + \sigma^2\,\frac{(x_* - \boldsymbol{w}'\boldsymbol{W}^{-1}\boldsymbol{X})^2}{\boldsymbol{X}'\boldsymbol{W}^{-1}\boldsymbol{X}} \,.$$

Hence we deduce that

$$R(\hat{p}_1^\Delta) \leqq f(\Delta) = \sigma_*^2 - \sigma^2\boldsymbol{w}'\boldsymbol{W}^{-1}\boldsymbol{w} + \sigma^2\,\frac{\Delta^2(x_* - \boldsymbol{w}'\boldsymbol{W}^{-1}\boldsymbol{X})^2}{1 + \Delta^2\boldsymbol{X}'\boldsymbol{W}^{-1}\boldsymbol{X}} \leqq R(\hat{p}_3) \,.$$

5.5 When a faulty assumption is better than none at all

In the above development of improved estimators and predictors, we have continually encountered the problem of nuisance parameters. That is, the optimal technique very often depends upon certain unknown values. Section 5.4 suggested one way out when bounds for these values are known. We now consider the implication of substituting a fixed but faulty value in place of the true parameter. As we shall see, in several cases such a faulty assumption could be better than none at all. For a similar notion in the context of linear regression constraints see TORO-VISZCARRONDO and WALLACE (1968) and WALLACE and TORO-VISZCARRONDO (1969).

Consider for instance the optimal homogeneous predictor $\hat{\boldsymbol{p}}_2$ given by (5.2.16). This depends upon $\sigma^{-1}\boldsymbol{\beta}$, so let us call it $\hat{\boldsymbol{p}}_2(\sigma^{-1}\boldsymbol{\beta})$. Now what, we might ask, would be the properties of $\hat{\boldsymbol{p}}_2(\boldsymbol{a})$ defined as

$$\hat{\boldsymbol{p}}_2(\boldsymbol{a}) = (\boldsymbol{X}^*\boldsymbol{a}\boldsymbol{a}'\boldsymbol{X} + \boldsymbol{W}_0)'\,(\boldsymbol{X}\boldsymbol{a}\boldsymbol{a}'\boldsymbol{X}' + \boldsymbol{W})^{-1}\,\boldsymbol{y} \tag{5.5.1}$$

obtained by substituting $\boldsymbol{a}$ for $\sigma^{-1}\boldsymbol{\beta}$ in (5.2.16). Clearly

$$R(\hat{\boldsymbol{p}}_2(\boldsymbol{a})) \geqq R(\hat{\boldsymbol{p}}_2(\sigma^{-1}\boldsymbol{\beta})) = \min_{\boldsymbol{a}\in E^p} R(\hat{\boldsymbol{p}}_2(\boldsymbol{a}))\,,$$

by virtue of the definition of R-optimality. Now we are particularly interested in the set $\mathfrak{A}$ of vectors $\boldsymbol{a}$ which are such that $\hat{\boldsymbol{p}}_2(\boldsymbol{a})$ is an improvement over the GAUSS-MARKOV predictor $\hat{\boldsymbol{p}}_3$. For this to be so it is necessary that

$$f(\boldsymbol{a}) = R(\hat{\boldsymbol{p}}_3) - R(\boldsymbol{p}_2(\boldsymbol{a})) \tag{5.5.2}$$

should be non-negative. TOUTENBURG (1975, pp. 82—83) derives sufficient conditions for this to be so. He also applies similar considerations to the problem of estimation (TOUTENBURG 1975, p. 84), and investigates this in general. A special case of the 'improvement region' which he derives has already been considered in (2.1.12).

5.6 Qualitative assertions for random a

The vector $\sigma^{-1}\boldsymbol{\beta}$ in $\hat{\boldsymbol{p}}_2(\sigma^{-1}\boldsymbol{\beta})$ can in certain cases be estimated. In such cases the risk function $R(\hat{\boldsymbol{p}}_2(\boldsymbol{a}))$ will be only approximately accurate. We now investigate the degree of this approximation for a particularly simple model.

Suppose that

$$y_t = \mu + u_t\,, \qquad \boldsymbol{u} \sim N(\boldsymbol{0}, \boldsymbol{I})\,. \tag{5.6.1}$$

The unbiased homogeneous estimator of μ is the GAUSS-MARKOV and everyday estimator $\hat{\beta}_3 = \bar{y}$ with

$$R(\hat{\beta}_3) = E(\bar{y} - \mu)^2 = n^{-1}\,. \tag{5.6.2}$$

The R-optimal homogeneous estimator derived as in Section 4.3 is

$$\hat{\beta}_2(\mu) = \frac{n\mu^2}{1 + n\mu^2}\,\bar{y}\,. \tag{5.6.3}$$

We may substitute for the unknown parameter μ in (5.6.3) using the estimator $\bar{y}$. This gives the function

$$\beta_2(\bar{y}) = \frac{n\bar{y}^3}{1 + ny^2} = h(\bar{y})$$

which has already been encountered in (2.4.1). The first few moments of this function can be approximated by expanding $h(.)$ in a TAYLOR series around the value $\bar{y} = \mu$. This gives the following approximations:

$$E\hat{\beta}_2(\bar{y}) = h(\mu) + \frac{1}{2}\frac{\partial^2 h}{\partial \bar{y}^2}(\mu)\, E(\bar{y} - \mu)^2 ,$$

and

$$var\,\hat{\beta}_2(\bar{y}) = E\left[\frac{\partial h}{\partial \bar{y}}(\mu)\,(\bar{y} - \mu) + \frac{1}{6}\frac{\partial^3 h}{\partial \bar{y}^3}(\mu)\,(\bar{y} - \mu)^3 + \frac{1}{2}\frac{\partial^2 h}{\partial \bar{y}^2}(\mu)\,\{(\bar{y} - \mu)^2 - E(\bar{y} - \mu)^2\}\right]^2 .$$

Using the abbreviation $z = n\mu^2$ these expressions simplify to

$$E\hat{\beta}_2(\bar{y}) = (1 + z)^{-3}\,(3z + 2z^3 + z^4)\,\mu$$

and

$$var\,\hat{\beta}_2(\bar{y}) = (1 + z)^{-8}\,(z^8 + 10z^7 + 45z^6 + 72z^5 - 46z^4 - 346z^3 + 537z^2 - 144z + 15)\,n^{-1} .$$

The risk of $\hat{\beta}_2(\bar{y})$ can be approximated using these two expressions. Within the region of approximation we may show that $\hat{\beta}_2(\bar{y})$ has a lower risk than $\bar{y}$ when and only when

$$g(z) = -3z^7 - 26z^6 - 34z^5 + 112z^4 + 417z^3 - 505z^2 + 148z - 14 \geqq 0 .$$

This condition is fulfilled for certain regions (z_1, z_2). This way the modified estimator $\hat{p}_2(\bar{y})$ has a smaller risk than the OLS when and only when

$$\mu \in \left(\sqrt{\frac{z_1}{n}},\ \sqrt{\frac{z_2}{n}}\right) .$$

The limits of these intervals depend on the degree of approximation given by the TAYLOR series. They tend to zero if n goes to infinity.

These results are only of a qualitative nature which indicates the limits of our proposed method.

5.7 Summary

This chapter develops the methods of estimation advanced in Chapter 4, and applies them in the context of prediction. The restrictive nature of the conventional definition of unbiased prediction, given in (5.1.5), is criticized.

The three standard types of optimal predictors are derived in Section 5.2. The optimal heterogeneous predictor and the optimal homogeneous unbiased predictor are related to methods proposed by GOLDBERGER (1962).

In Section 5.3 we relate these results to those derived in Chapter 4, and calculate the gains in efficiency that may be made by their use.

Section 5.4 looks at the effect of auxiliary inequalities. An example is given using the COBB-DOUGLAS production function, and the special case of a single regressor variable is examined in some detail.

The final sections examine when a faulty assumption may be better than no assumption at all, and also use TAYLOR approximations to investigate the properties of randomized methods based on the above procedures.

6

Prediction with linear constraints

6.1 Examples of linear constraints

Sometimes in addition to having direct observations on the explanatory and dependent variables we also have auxiliary information on the vector of regression coefficients. When this takes the form of linear inequalities, various types of simplex algorithm can be used to find a numerical solution. However this chapter looks at auxiliary information which can be written as linear equalities in the form

$$\boldsymbol{r} = \boldsymbol{R\beta} + \boldsymbol{d} . \tag{6.1.1}$$

Here $\boldsymbol{r}$ and $\boldsymbol{R}$ are known (of order $J \times 1$ and $J \times p$ respectively, with rank $R = J$), and $\boldsymbol{d}$ is a vector of random disturbances. We assume that $\boldsymbol{d}$ satisfies

$$\boldsymbol{d} \sim (\boldsymbol{0}, \boldsymbol{V}) , \quad \boldsymbol{V} \text{ positive definite}, \quad E[\boldsymbol{du}'] = \boldsymbol{0} , \quad E[\boldsymbol{d}u_*] = \boldsymbol{0} . \tag{6.1.2}$$

Several examples of such a situation spring to mind.

(a) A separate estimate of $\boldsymbol{\beta}$ might exist say from an earlier sample. If this estimate is called $\boldsymbol{b}^*$ than we might have

$$\boldsymbol{b}^* = \boldsymbol{\beta} + \boldsymbol{d}$$

which is clearly a special case of (6.1.1).

(b) Another special case occurs when $\boldsymbol{V}$, the dispersion matrix of $\boldsymbol{d}$, is null. This corresponds to knowing that $\boldsymbol{r}$ equals $\boldsymbol{R\beta}$ with certainty. (For instance we might know that two coefficients are equal.)

(c) A third special case arises when particular elements of $\boldsymbol{\beta}$ are known. In this case $\boldsymbol{R}$ equals $[\boldsymbol{I}, \boldsymbol{0}]$, or some other matrix of zeros and ones, with ones corresponding to the elements of $\boldsymbol{\beta}$ that are known.

(d) Alternatively there may be information concerning the ratios between certain coefficients. For instance, if the ratio $\beta_1 : \beta_2 : \beta_3$ is $ab : b : 1$, this may be written

$$\begin{bmatrix} 1 & -a & 0 \\ 0 & 1 & -b \\ 1 & 0 & -ab \end{bmatrix} \begin{bmatrix} \beta_1 \\ \beta_2 \\ \beta_3 \end{bmatrix} = \boldsymbol{0} .$$

Alternatively, the righthand side of this equation could be nonzero if there was some uncertainty about the ratios.

(e) Another possibility is if say the sum of the coefficients is known. This may occur in the well-known COBB-DOUGLAS production function with constant returns to scale, for which $\beta_1 + \beta_2 = 1$ in (5.4.12).

(f) Finally we may have reason to believe that certain elements of $\boldsymbol{\beta}$ lie in a particular region. Suppose for instance that β_i is known to lie between a and b. Then we can put

$$\beta_i = \tfrac{1}{2}(a + b) + d\,,$$

where d might be uniformly distributed on the range $\pm \tfrac{1}{2}(a - b)$. (See THEIL (1963) and TOUTENBURG (1975, p. 89).)

All these possibilities, as well as many others, can be captured by the general specification (6.1.1), and will be investigated in the succeeding sections.

6.2 Stochastic constraints

6.2.1 Estimation and prediction. We now return to the general situation represented by (6.1.1). Putting this equation alongside (5.1.1) gives

$$\begin{bmatrix} \boldsymbol{y} \\ \boldsymbol{r} \end{bmatrix} = \begin{bmatrix} \boldsymbol{X} \\ \boldsymbol{R} \end{bmatrix} \beta + \begin{bmatrix} \boldsymbol{u} \\ \boldsymbol{d} \end{bmatrix}. \tag{6.2.1}$$

Calling the augmented matrices $\tilde{\boldsymbol{y}}$, $\tilde{\boldsymbol{X}}$ and $\tilde{\boldsymbol{u}}$ this can be written

$$\tilde{\boldsymbol{y}} = \tilde{\boldsymbol{X}}\beta + \tilde{\boldsymbol{u}}\,.$$

Assuming that the elements of $\boldsymbol{u}$ are uncorrelated with the elements of $\boldsymbol{d}$, we know that $\tilde{\boldsymbol{u}} \sim (\boldsymbol{0}, \sigma^2\boldsymbol{\varphi})$ where

$$\boldsymbol{\varphi} = \begin{bmatrix} \boldsymbol{W} & \boldsymbol{0} \\ \boldsymbol{0} & \sigma^{-2}\boldsymbol{V} \end{bmatrix}. \tag{6.2.2}$$

Since $\boldsymbol{W}$ and $\boldsymbol{V}$ are positive definite we know that $\boldsymbol{\varphi}$ is also.

Hence stochastic constraints like (6.1.1) may easily be incorporated within the general linear model. (If the elements of $\boldsymbol{u}$ and $\boldsymbol{d}$ are correlated the only effect is to change the off-diagonal sub-matrices of $\boldsymbol{\varphi}$.)

It would seem superfluous to repeat here all the procedures outlined in Chapter 5 for the situation with linear restrictions. Hence we limit ourselves to a single example — the development of the unbiased homogeneous predictor with restrictions.

We seek a predictor of the form

$$\boldsymbol{p} = \boldsymbol{C}'\tilde{\boldsymbol{y}}\,. \tag{6.2.3}$$

(That is, $\boldsymbol{p}$ depends not only on $\boldsymbol{y}$, but also on the auxiliary vector $\boldsymbol{r}$.)

As in (5.2.17) the condition of unbiasedness requires that

$$C'\tilde{X} = X_* . \qquad (6.2.4)$$

We also need the fact that

$$E[uu'_*] = \sigma^2 \begin{bmatrix} W_0 \\ 0 \end{bmatrix} = \sigma^2\psi, \text{ say.} \qquad (6.2.5)$$

Taking as loss function

$$R(p) = E[(p - y_*)' A(p - y_*)] \qquad (6.2.6)$$

we see that

$$R(p) = \sigma^2 \, tr\, A(C'\varphi C + W_* - 2C'\psi) . \qquad (6.2.7)$$

To minimize this subject to the constraints given by (6.2.4), we add a matrix of LAGRANGIAN multipliers, as in (5.2.19). After some simplification this leads to the following results, which were obtained by THEIL and GOLDBERGER (1961), and KAKWANI (1965), respectively.

Theorem 6.1 (a) The optimal unbiased linear estimator in the model given by (6.2.2) is

$$\hat{\beta}_4 = (\sigma^{-2}S + R'V^{-1}R)^{-1} (\sigma^{-2}X'W^{-1}y + R'V^{-1}r) . \qquad (6.2.8)$$

The dispersion matrix of $\hat{\beta}_4$ is

$$V_{\hat{\beta}_4} = V_4 = \sigma^2(S + \sigma^2 R'V^{-1}R)^{-1} . \qquad (6.2.9)$$

(b) The optimal unbiased linear predictor in the model given by (6.2.2) is

$$\hat{p}_4 = X_*\hat{\beta}_4 + W_0'W^{-1}(y - X\hat{\beta}_4) , \qquad (6.2.10)$$

where $\hat{\beta}_4$ is given in (6.2.8). The value of its loss function is

$$R(\hat{p}_4) = R(\hat{p}_1) + tr\, A(X_* - W_0'W^{-1}X)\, V_4(X_* - W_0'W^{-1}X)' ,$$

where $R(\hat{p}_1)$ is given by (5.2.10) and V_4 by (6.2.9).

Finally we note the special case where multicollinearity in X is accounted for by the equation $R\beta = 0$, a special case of (6.1.1). Putting $r = d = 0$ and allowing V to tend to zero in (6.2.8) gives the following.

Theorem 6.2 When $y = X\beta + u$ where $u \sim (0, \sigma^2 W)$, *rank* $X = p_1 < p$ and $R\beta = 0$, *rank* $R = p - p_1$, *rank* $(X, R) = p$, then the optimal unbiased linear estimator of β is

$$\hat{\beta}(R) = (S + R'R)^{-1} X'W^{-1}y . \qquad (6.2.11)$$

This has dispersion matrix

$$V_{\hat{\beta}(R)} = \sigma^2(S + R'R)^{-1} S(S + R'R)^{-1} . \qquad (6.2.12)$$

The estimator defined in (6.2.11) has a close formal similarity with the *ridge estimator* defined by HOERL and KENNARD (1970). The estimator

$\hat{\beta}(\boldsymbol{R})$ is the best conditional unbiased estimator in the model defined by the theorem's assumptions.

6.2.2 Gains in efficiency. One of the main effects of auxiliary information such as (6.1.1) is to reduce the dispersion matrix of the optimal estimator. Equation (6.2.9) may be compared with the matrix $\boldsymbol{V}_3 = \sigma^2\boldsymbol{S}^{-1}$ which gave the dispersion of the generalized least squares estimator. Note that

$$\boldsymbol{V}_4^{-1} - \boldsymbol{V}_3^{-1} = \boldsymbol{R}'\boldsymbol{V}^{-1}\boldsymbol{R} \tag{6.2.13}$$

and this matrix is certainly nonnegative definite (see Theorem A8). Alternatively we may examine

$$\sigma^{-2}(\boldsymbol{V}_4 - \boldsymbol{V}_3) = (\boldsymbol{S} + \sigma^2\boldsymbol{R}'\boldsymbol{V}^{-1}\boldsymbol{R})^{-1} - \boldsymbol{S}^{-1}\,.$$

This matrix is nonpositive definite (see Theorem A 11), and measures the extent by which the information contained in (6.1.1) has 'reduced' the dispersion of the optimal estimator. A similar result follows for the predictor defined by (6.2.10) according to

$$R(\hat{\boldsymbol{p}}_4) - R(\hat{\boldsymbol{p}}_3) = tr\,\boldsymbol{A}(\boldsymbol{X}_* - \boldsymbol{W}_0'\boldsymbol{W}^{-1}\boldsymbol{X})\,(\boldsymbol{V}_4 - \boldsymbol{V}_3)\,(\boldsymbol{X}_* - \boldsymbol{W}_0'\boldsymbol{W}^{-1}\boldsymbol{X})' \leqq 0\,.$$

See also KAKWANI (1965, p. 103).

6.3 Auxiliary information on σ^2

Since $\hat{\boldsymbol{\beta}}_4$ given by (6.2.8) depends on σ^2, which in general is unknown, the estimator $\hat{\boldsymbol{\beta}}_4$ is in many cases not practicable. Only in special cases, such as when σ^2 is known from a preliminary sample, can the formulae given in (6.2.8) and (6.2.10) be evaluated. (An alternative special case occurs when $\boldsymbol{V} = \sigma^2\boldsymbol{V}^*$ and $\boldsymbol{V}^*$ is known).

However, (6.2.8) is just one member of the family of unbiased estimators

$$\hat{\boldsymbol{\beta}}_c = (c\boldsymbol{S} + \boldsymbol{R}'\boldsymbol{V}^{-1}\boldsymbol{R})^{-1}\,(c\boldsymbol{X}'\boldsymbol{W}^{-1}\boldsymbol{y} + \boldsymbol{R}'\boldsymbol{V}^{-1}\boldsymbol{r}) \tag{6.3.1}$$

whose members are indexed by the value of the nonnegative scalar c. Note that the matrix

$$\boldsymbol{M}_c = c\boldsymbol{S} + \boldsymbol{R}'\boldsymbol{V}^{-1}\boldsymbol{R} \tag{6.3.2}$$

is positive definite for all positive c. Also since

$$\boldsymbol{X}'\boldsymbol{W}^{-1}\boldsymbol{y} = \boldsymbol{X}'\boldsymbol{W}^{-1}(\boldsymbol{X}\boldsymbol{\beta} + \boldsymbol{u}) = \boldsymbol{S}\boldsymbol{\beta} + \boldsymbol{X}'\boldsymbol{W}^{-1}\boldsymbol{u}$$

and

$$\boldsymbol{R}'\boldsymbol{V}^{-1}\boldsymbol{r} = \boldsymbol{R}'\boldsymbol{V}^{-1}(\boldsymbol{R}\boldsymbol{\beta} + \boldsymbol{d}) = \boldsymbol{R}'\boldsymbol{V}^{-1}\boldsymbol{R}\boldsymbol{\beta} + \boldsymbol{R}'\boldsymbol{V}^{-1}\boldsymbol{d}\,,$$

the term in the final brackets of (6.3.1) equals

$$\boldsymbol{M}_c\boldsymbol{\beta} + \boldsymbol{X}'\boldsymbol{W}^{-1}\boldsymbol{u} + \boldsymbol{R}'\boldsymbol{V}^{-1}\boldsymbol{d}\,.$$

Hence on simplifying, (6.3.1) becomes

$$\begin{aligned}\hat{\boldsymbol{\beta}}_c &= \boldsymbol{M}_c^{-1}(\boldsymbol{M}_c\boldsymbol{\beta} + c\boldsymbol{X}'\boldsymbol{W}^{-1}\boldsymbol{u} + \boldsymbol{R}'\boldsymbol{V}^{-1}\boldsymbol{d}) \\ &= \boldsymbol{\beta} + \boldsymbol{M}_c^{-1}(c\boldsymbol{X}'\boldsymbol{W}^{-1}\boldsymbol{u} + \boldsymbol{R}'\boldsymbol{V}^{-1}\boldsymbol{d})\,.\end{aligned}$$

From this it may be shown (see also (6.1.2)) that $\hat{\boldsymbol{\beta}}_c$ is unbiased for all c, and has the dispersion matrix

$$\boldsymbol{V}_c = \boldsymbol{M}_c^{-1}(c^2\sigma^2\boldsymbol{S} + \boldsymbol{R}'\boldsymbol{V}^{-1}\boldsymbol{R})\,\boldsymbol{M}_c^{-1}\,.$$

Equation (6.2.9) is of course just a special case of this.

It is also possible to construct a predictor based on $\hat{\boldsymbol{\beta}}_c$. Consider

$$\hat{\boldsymbol{p}}_c = \boldsymbol{X}_*\hat{\boldsymbol{\beta}}_c + \boldsymbol{W}_0'\boldsymbol{W}^{-1}(\boldsymbol{y} - \boldsymbol{X}\hat{\boldsymbol{\beta}}_c)\,.$$

This is a generalization of (6.2.10), and has the properties

$$E[\hat{\boldsymbol{p}}_c - \boldsymbol{y}_*] = \boldsymbol{0}\,,$$

$$R(\hat{\boldsymbol{p}}_c) = R(\hat{\boldsymbol{p}}_1) + tr\,\boldsymbol{A}(\boldsymbol{X}_* - \boldsymbol{W}_0'\boldsymbol{W}^{-1}\boldsymbol{X})\,\boldsymbol{V}_c(\boldsymbol{X}_* - \boldsymbol{W}_0'\boldsymbol{W}^{-1}\boldsymbol{X})'\,.$$

Because of the optimality of $\hat{\boldsymbol{\beta}}_{\sigma^{-2}} = \hat{\boldsymbol{\beta}}_4$ we deduce that

$$R(\hat{\boldsymbol{p}}_c) - R(\hat{\boldsymbol{p}}_4) \geqq 0\,.$$

Of particular interest is the limiting value of $\hat{\boldsymbol{\beta}}_c$ when c tends to zero and when c tends to infinity. We may deduce that

$$\hat{\boldsymbol{\beta}}_\infty = \lim_{c\to\infty} \hat{\boldsymbol{\beta}}_c = \boldsymbol{S}^{-1}\boldsymbol{X}'\boldsymbol{W}^{-1}\boldsymbol{y} = \boldsymbol{b}$$

$$\hat{\boldsymbol{\beta}}_0 = \lim_{c\to 0} \hat{\boldsymbol{\beta}}_c = (\boldsymbol{R}'\boldsymbol{V}^{-1}\boldsymbol{R})^{-1}\,\boldsymbol{R}'\boldsymbol{V}^{-1}\boldsymbol{r}\,.$$

The replacement of σ^{-2} in $\hat{\boldsymbol{\beta}}_4$ by a nonstochastic value c is useful only where the resulting estimator $\hat{\boldsymbol{\beta}}_c$ has smaller risk than the GLSE. If we have auxiliary information on σ such that $\sigma^2 > \sigma_1^2$, then putting $c = \sigma_1^{-2}$ in (6.3.1) gives an estimator $\hat{\boldsymbol{\beta}}_{\sigma_1^{-2}}$ which has smaller risk than the GLSE.

6.4 Exact constraints

Section 6.2 considered stochastic constraints of the form $\boldsymbol{r} = \boldsymbol{R}\boldsymbol{\beta} + \boldsymbol{d}$, where $\boldsymbol{d}$ was random with mean zero and dispersion matrix $\boldsymbol{V}$. As rank $\boldsymbol{V} < J$ and $\boldsymbol{V}$ tends to the null matrix, these stochastic constraints tend to the exact constraint $\boldsymbol{r} = \boldsymbol{R}\boldsymbol{\beta}$. For a maximum likelihood approach to estimation under such conditions the reader is referred to MARDIA and BIBBY (1977). However the problem can also be approached, following

TOUTENBURG (1975, p. 98), by assuming rank $V < J$ and considering a particular form of heterogeneous estimator, namely

$$p = C_1'y + C_2'r . \tag{6.4.1}$$

Since

$$p - y_* = (C_1'X + C_2'R - X_*)\,\beta + (C_1'u + C_2'd - u_*)\,,$$

it follows that p is unbiased for y_* only when

$$C_1'X + C_2'R - X_* = 0 . \tag{6.4.2}$$

Hence for the optimal estimator of the form given by (6.4.1) we must minimize

$$R(p) = tr\,A(\sigma^2 C_1'WC_1 + C_2'VC_2 + \sigma^2 W_* - 2\sigma^2 C_1'W_0)$$

subject to the constraints given by (6.4.2). Inserting a *Lagrangian* matrix Λ and differentiating leads to the following normal equations:

$$\sigma^2 AC_1'W - \sigma^2 AW_0 - \Lambda X' = 0\,,$$

$$AC_2V \quad - \Lambda R' = 0$$

and

$$C_1'X + C_2'R - X_* = 0 .$$

These equations have the unique solution

$$\hat{p}_5 = \hat{C}_1'y + \hat{C}_2'r = X_* b_5 + W_0'W^{-1}(y - Xb_5)\,, \tag{6.4.3}$$

where

$$b_5 = b + S^{-1}R'(\sigma^{-2}V + RS^{-1}R')^{-1}\,(r - Rb) \tag{6.4.4}$$

and b is the generalized least squares estimator. These estimators are unbiased when $R\beta = r$, and their loss function and dispersion matrices under these conditions are as follows:

$$R(\hat{p}_5) = R(\hat{p}_1) + tr\,A(X_* - W_0'W^{-1}X)\,V_5(X_* - W_0'W^{-1}X)'\,, \tag{6.4.5}$$

where

$$V_5 = \sigma^2 S^{-1} - \sigma^2 S^{-1}R'(\sigma^{-2}V + RS^{-1}R')^{-1}RS^{-1} . \tag{6.4.6}$$

Note that V_5, the dispersion matrix of b_5, is 'less' than that of the generalized least squares vector, by the amount of the last term in (6.4.6), and also that the risk of $\hat{p}_5$ given by (6.4.5) exceeds that of $\hat{p}_1$ given by (5.2.10).

The following GAUSS-MARKOV property concerning b_5 can easily be proved.

Theorem 6.3. The estimator b_5 defined in (6.4.4) is the BLUE in the class of heterogeneous estimators of the form

$$\hat{\beta} = D_1 y + D_2 r .$$

Proof. See TOUTENBURG (1975, p. 99).

6.5 Auxiliary estimates and piecewise regression

6.5.1 Definition of the procedure. We now turn to consider the estimation of a vector, for part of which a prior estimator also exists. That is, the vector $\boldsymbol{\beta}$ is of interest, where $\boldsymbol{\beta}' = (\boldsymbol{\beta}_1', \boldsymbol{\beta}_2')$, and the subvector $\boldsymbol{\beta}_1$ has a prior estimator $\boldsymbol{b}_1^*$ which satisfies

$$\boldsymbol{b}_1^* = \boldsymbol{\beta}_1 + \boldsymbol{d} . \tag{6.5.1}$$

Toutenburg (1975, p. 100) calls this procedure "*schrittweise Regression*" which is literally translated as "*stepwise regression*". However since this term has a different connotation in the English literature, we shall use the less accurate translation '*piecewise regression*'. Suppose then that

$$\boldsymbol{y} = \boldsymbol{X}_1\boldsymbol{\beta}_1 + \boldsymbol{X}_2\boldsymbol{\beta}_2 + \boldsymbol{u} . \tag{6.5.2}$$

We use $\boldsymbol{y}_-$ to denote the residual vector obtained after fitting $\boldsymbol{y}$ by the prior estimate $\boldsymbol{b}_1^*$. In other words

$$\boldsymbol{y}_- = \boldsymbol{y} - \boldsymbol{X}_1\boldsymbol{b}_1^* = \boldsymbol{X}_1(\boldsymbol{\beta}_1 - \boldsymbol{b}_1^*) + \boldsymbol{X}_2\boldsymbol{\beta}_2 + \boldsymbol{u} .$$

Using (6.5.1) this equals

$$\boldsymbol{y}_- = \boldsymbol{X}_2\boldsymbol{\beta}_2 + \boldsymbol{u} - \boldsymbol{X}_1\boldsymbol{d} = \boldsymbol{X}_2\boldsymbol{\beta}_2 + \tilde{\boldsymbol{u}} , \tag{6.5.3}$$

where $\tilde{\boldsymbol{u}} = \boldsymbol{u} - \boldsymbol{X}_1\boldsymbol{d}$.

Now from the assumptions given in (6.1.2), the vector $\tilde{\boldsymbol{u}}$ has mean zero and dispersion matrix

$$E[\tilde{\boldsymbol{u}}\tilde{\boldsymbol{u}}'] = \sigma^2\boldsymbol{W} + \boldsymbol{X}_1\boldsymbol{V}\boldsymbol{X}_1' = \widetilde{\boldsymbol{W}} ,$$

say. Furthermore, we define

$$\tilde{\boldsymbol{u}}_* = \boldsymbol{u}_* - \boldsymbol{X}_{1*}\boldsymbol{d} ,$$

which has the following moments

$$E[\tilde{\boldsymbol{u}}_*\tilde{\boldsymbol{u}}_*'] = \widetilde{\boldsymbol{W}}_* = \sigma^2\boldsymbol{W}_* + \boldsymbol{X}_{1*}\boldsymbol{V}\boldsymbol{X}_{1*}' ,$$

and

$$E[\boldsymbol{u}_*\tilde{\boldsymbol{u}}_*'] = \widetilde{\boldsymbol{W}}_0 = \sigma^2\boldsymbol{W}_0 + \boldsymbol{X}_1\boldsymbol{V}\boldsymbol{X}_1' .$$

Hence the generalized least squares estimator of $\boldsymbol{\beta}_2$ in the model (6.5.3) is

$$\boldsymbol{b}_2^* = (\boldsymbol{X}_2'\widetilde{\boldsymbol{W}}^{-1}\boldsymbol{X}_2)^{-1}\,\boldsymbol{X}_2'\widetilde{\boldsymbol{W}}^{-1}\boldsymbol{y}_- .$$

This has dispersion matrix

$$\boldsymbol{V}(\boldsymbol{b}_2^*) = (\boldsymbol{X}_2'\widetilde{\boldsymbol{W}}^{-1}\boldsymbol{X}_2)^{-1}$$

and leads to the predictor

$$\hat{\boldsymbol{p}}_6 = \boldsymbol{X}_{1*}\boldsymbol{b}_1^* + \boldsymbol{X}_{2*}\boldsymbol{b}_2^* + \widetilde{\boldsymbol{W}}_0'\widetilde{\boldsymbol{W}}^{-1}(\boldsymbol{y}_- - \boldsymbol{X}_2\boldsymbol{b}_2^*) . \tag{6.5.4}$$

We calculate the risk of $\hat{\boldsymbol{p}}_6$ as follows.

$$\hat{\boldsymbol{p}}_6 - \boldsymbol{y}_* = (\boldsymbol{X}_{2*} - \widetilde{\boldsymbol{W}}_0'\widetilde{\boldsymbol{W}}^{-1}\boldsymbol{X}_2)\,\boldsymbol{V}(\boldsymbol{b}_2^*)\,\boldsymbol{X}_2'\widetilde{\boldsymbol{W}}^{-1}\tilde{\boldsymbol{u}} - \tilde{\boldsymbol{u}}_* + \widetilde{\boldsymbol{W}}_0'\widetilde{\boldsymbol{W}}^{-1}\tilde{\boldsymbol{u}} .$$

Therefore

$$R(\hat{\boldsymbol{p}}_6) = tr\,\boldsymbol{A}\Big[(\boldsymbol{X}_* - \widetilde{\boldsymbol{W}}_0'\widetilde{\boldsymbol{W}}^{-1}[\boldsymbol{0}, \boldsymbol{X}_2])\begin{pmatrix}\boldsymbol{V} & \boldsymbol{0}\\ \boldsymbol{0} & \boldsymbol{V}(\boldsymbol{b}_2^*)\end{pmatrix}$$
$$\times (\boldsymbol{X}_* - \widetilde{\boldsymbol{W}}_0'\widetilde{\boldsymbol{W}}^{-1}[\boldsymbol{0}, \boldsymbol{X}_2])'\Big] + tr\,\boldsymbol{A}(\sigma^2\widetilde{\boldsymbol{W}}_* - \widetilde{\boldsymbol{W}}_0'\widetilde{\boldsymbol{W}}^{-1}\widetilde{\boldsymbol{W}}_0)\,. \tag{6.5.5}$$

6.5.2 Gains in efficiency. This section compares the efficiencies of the various estimators derived above. The results are summarized in Theorem 6.4 at the end of the section, and readers who are concerned with results rather than proofs may choose to omit the intervening pages. The predictor $\hat{\boldsymbol{p}}_6$ given by (6.5.4) may be compared with the unconstrained GLS predictor $\hat{\boldsymbol{p}}_3$ defined in (5.2.22), and with the constrained predictors $\hat{\boldsymbol{p}}_4$ and $\hat{\boldsymbol{p}}_5$ from (6.2.10) and (6.4.3) respectively.

a) If rank $\boldsymbol{V} = J$, then as $\hat{\boldsymbol{p}}_4$ is the R-optimal predictor we immediately have

$$R(\hat{\boldsymbol{p}}_6) - R(\hat{\boldsymbol{p}}_4) \geqq 0\,.$$

b) If rank $\boldsymbol{V} < J$, then we must distinguish between exact and inexact prior knowledge of the subvector $\boldsymbol{\beta}_1$. In the first case we have $\boldsymbol{V} = \boldsymbol{0}$ and therefore $\widetilde{\boldsymbol{W}} = \sigma^2\boldsymbol{W}$, $\widetilde{\boldsymbol{W}}_* = \sigma^2\boldsymbol{W}_*$, $\widetilde{\boldsymbol{W}}_0 = \sigma^2\boldsymbol{W}_0$. Putting $\boldsymbol{V} = \boldsymbol{0}$ in equation (6.5.4) we obtain

$$\hat{\boldsymbol{p}}_6 = \boldsymbol{X}_{1*}\boldsymbol{\beta}_1 + \boldsymbol{X}_{2*}\boldsymbol{b}_2^* + \boldsymbol{W}_0'\boldsymbol{W}^{-1}(\boldsymbol{y} - \boldsymbol{X}_2\boldsymbol{b}_2^*)\,,$$

where

$$\boldsymbol{b}_2^* = \boldsymbol{S}_{22}^{-1}\boldsymbol{X}_2'\boldsymbol{W}^{-1}\boldsymbol{y}_-\,.$$

Now

$$\boldsymbol{b}_2^* - \boldsymbol{\beta}_2 = \boldsymbol{S}_{22}^{-1}\boldsymbol{X}_2'\boldsymbol{W}^{-1}\boldsymbol{u}\,, \tag{6.5.6}$$

and

$$\boldsymbol{V}(\boldsymbol{b}_2^*) = \sigma^2\boldsymbol{S}_{22}^{-1} = \sigma^2(\boldsymbol{X}_2'\boldsymbol{W}^{-1}\boldsymbol{X}_2)^{-1}\,.$$

Therefore

$$\boldsymbol{b}_6 = \begin{pmatrix}\boldsymbol{\beta}_1\\ \boldsymbol{b}_2^*\end{pmatrix}, \quad \text{and} \quad \boldsymbol{V}_6 = \begin{pmatrix}\boldsymbol{0} & \boldsymbol{0}\\ \boldsymbol{0} & \boldsymbol{V}(\boldsymbol{b}_2^*)\end{pmatrix}. \tag{6.5.7}$$

Finally we have

$$R(\hat{\boldsymbol{p}}_6) = R(\hat{\boldsymbol{p}}_1) + tr\,\boldsymbol{A}[(\boldsymbol{X}_* - \boldsymbol{W}_0'\boldsymbol{W}^{-1}\boldsymbol{X})\,\boldsymbol{V}_6(\boldsymbol{X}_* - \boldsymbol{W}_0'\boldsymbol{W}^{-1}\boldsymbol{X})']\,. \tag{6.5.8}$$

In comparing the predictors $\hat{\boldsymbol{p}}_6$ and $\hat{\boldsymbol{p}}_3$ we need the components $\boldsymbol{b}_1$, $\boldsymbol{b}_2$ of $\boldsymbol{b}$ which correspond to the components $\boldsymbol{\beta}_1$, $\boldsymbol{\beta}_2$ of $\boldsymbol{\beta}$. If we denote $\boldsymbol{S}_{ij} = \boldsymbol{X}_i'\boldsymbol{W}^{-1}\boldsymbol{X}_j$ $(i, j = 1, 2)$ and use (6.5.2) we get

$$\boldsymbol{b} = \begin{pmatrix}\boldsymbol{b}_1\\ \boldsymbol{b}_2\end{pmatrix} = \begin{pmatrix}\boldsymbol{S}_{11} & \boldsymbol{S}_{12}\\ \boldsymbol{S}_{21} & \boldsymbol{S}_{22}\end{pmatrix}^{-1}\begin{pmatrix}\boldsymbol{X}_1'\;\boldsymbol{W}^{-1}\boldsymbol{y}\\ \boldsymbol{X}_2'\;\boldsymbol{W}^{-1}\boldsymbol{y}\end{pmatrix}$$

and therefore (see also Theorem A 36)

$$b_2 = D^{-1}EW^{-1}y\,, \tag{6.5.9}$$

and

$$b_1 = S_{11}^{-1}X_1'W^{-1}y - S_{11}^{-1}S_{12}b_2\,, \tag{6.5.10}$$

where

$$\left.\begin{aligned} &E = (X_2' - S_{21}S_{11}^{-1}X_1')\,, \quad D = EW^{-1}X_2\,,\\ &EW^{-1}X_1 = 0\,, \quad EW^{-1}E' = D\,. \end{aligned}\right\} \tag{6.5.11}$$

We now calculate the components of the dispersion matrix

$$V_b = \sigma^2 S^{-1} = \begin{pmatrix} V_1 & V_{12} \\ V_{21} & V_2 \end{pmatrix}.$$

Using (6.5.9) to (6.5.11) the following relationships hold

$$b_2 - \beta_2 = D^{-1}EW^{-1}u\,, \tag{6.5.12}$$

$$V_2 = E(b_2 - \beta_2)(b_2 - \beta_2)' = \sigma^2 D^{-1}\,, \tag{6.5.13}$$

$$b_1 - \beta_1 = S_{11}^{-1}[X_1' - S_{12}D^{-1}E]\,W^{-1}u\,, \tag{6.5.14}$$

$$V_1 = E(b_1 - \beta_1)(b_1 - \beta_1)' = \sigma^2 S_{11}^{-1} + V_{12}V_2^{-1}V_{21}\,, \tag{6.5.15}$$

$$V_{21}' = V_{12} = E(b_1 - \beta_1)(b_2 - \beta_2)' = -\sigma^2 S_{11}^{-1}S_{12}D^{-1}\,, \tag{6.5.16}$$

giving

$$V_b = \sigma^2 S^{-1} = \sigma^2 \begin{pmatrix} S_{11}^{-1} + S_{11}^{-1}S_{12}D^{-1}S_{21}S_{11}^{-1} & -S_{11}^{-1}S_{12}D^{-1} \\ -D^{-1}S_{21}S_{11}^{-1} & D^{-1} \end{pmatrix}. \tag{6.5.17}$$

The matrix $(b - b_6)(b - b_6)'$ is nonnegative definite for all u (see Theorem A 8).

Therefore using

$$b_2^* - \beta_2 = S_{22}^{-1}X_2'W^{-1}u$$

and equations (6.5.9) to (6.5.11) we may conclude that

$$E(b - b_6)(b - b_6)' = \begin{pmatrix} V_1 & V_{12} \\ V_{21} & V_2 - V(b_2^*) \end{pmatrix} = V_3 - V_6$$

is nonnegative definite.

Therefore we may say that when $V = 0$ the predictors $\hat{p}_3$ and $\hat{p}_6$ satisfy

$$\begin{aligned} \delta(0) &= R(\hat{p}_3) - R(\hat{p}_6) \\ &= tr\,A[(X_* - W_0'W^{-1}X)(V_3 - V_6)(X_* - W_0'W^{-1}X)'] \geqq 0\,. \end{aligned} \tag{6.5.18}$$

That is, the piecewise predictor $\hat{p}_6$ is more efficient than the GLS predictor $\hat{p}_3$ if a subvector of β is known exactly. This result is not surprising but it yields the basis for the following.

Let us define

$$\delta(\boldsymbol{V}) = R(\hat{\boldsymbol{p}}_3) - R(\hat{\boldsymbol{p}}_6)\,,$$

where $\delta(\boldsymbol{0})$ is given in (6.5.18). Then we may conclude that whenever $\delta(\boldsymbol{0}) > 0$, the inequality $\delta(\boldsymbol{V}) > 0$ is satisfied throughout a certain neighbourhood of $\boldsymbol{V} = \boldsymbol{0}$. This means that the component $\boldsymbol{\beta}_1$ of $\boldsymbol{\beta}$ must be known sufficiently well but not necessary exactly to make $\hat{\boldsymbol{p}}_6$ better than $\hat{\boldsymbol{p}}_3$. We may now compare $\hat{\boldsymbol{p}}_5$ and $\hat{\boldsymbol{p}}_3$ using the information (6.5.1) written as $\boldsymbol{r} = \boldsymbol{R\beta}$ with $\boldsymbol{r} = \boldsymbol{b}_1^*$ and $\boldsymbol{R} = (\boldsymbol{I}, \boldsymbol{0})$. This gives a special form of the estimator $\boldsymbol{b}_5$ (see 6.4.4) as

$$\boldsymbol{b}_5 = \boldsymbol{b} + \boldsymbol{S}^{-1}\begin{pmatrix}\boldsymbol{I}\\ \boldsymbol{0}\end{pmatrix}\left((\boldsymbol{I},\boldsymbol{0})\boldsymbol{S}^{-1}\begin{pmatrix}\boldsymbol{I}\\ \boldsymbol{\theta}\end{pmatrix}\right)^{-1}(\boldsymbol{I},\boldsymbol{0})\begin{pmatrix}\boldsymbol{\beta}_1 - \boldsymbol{b}_1\\ \boldsymbol{\beta}_2 - \boldsymbol{b}_2\end{pmatrix}, \tag{6.5.19}$$

$$\boldsymbol{b}_5 = \begin{pmatrix}\boldsymbol{\beta}_1\\ \boldsymbol{b}_2 - \boldsymbol{V}_{21}\boldsymbol{V}_1^{-1}(\boldsymbol{b}_1 - \boldsymbol{\beta}_1)\end{pmatrix}. \tag{6.5.20}$$

From (6.4.12) to (6.5.16) we deduce that

$$\boldsymbol{V}_5 = \begin{pmatrix}\boldsymbol{0} & \boldsymbol{0}\\ \boldsymbol{0} & \boldsymbol{V}_2 - \boldsymbol{V}_{21}\boldsymbol{V}_1^{-1}\boldsymbol{V}_{12}\end{pmatrix}.$$

From this we deduce the result already quoted in Section 6.4 that $\boldsymbol{b}_5$ is better than the GLS estimator $\boldsymbol{b}$, i.e. that

$$\boldsymbol{V}_3 - \boldsymbol{V}_5 = \begin{pmatrix}\boldsymbol{V}_1 & \boldsymbol{V}_{12}\\ \boldsymbol{V}_{21} & \boldsymbol{V}_{21}\boldsymbol{V}_1^{-1}\boldsymbol{V}_{12}\end{pmatrix}$$

is nonnegative definite. Thus we conclude that

$$R(\hat{\boldsymbol{p}}_3) - R(\hat{\boldsymbol{p}}_5) \geqq 0\,.$$

Comparing now $\hat{\boldsymbol{p}}_6$ and $\hat{\boldsymbol{p}}_5$, we deduce from (6.5.7) and (6.5.19) that

$$\boldsymbol{b}_6 - \boldsymbol{b}_5 = \begin{pmatrix}\boldsymbol{0}\\ (\boldsymbol{b}_2^* - \boldsymbol{\beta}_2) - (\boldsymbol{b}_2 - \boldsymbol{\beta}_2) + \boldsymbol{V}_{21}\boldsymbol{V}_1^{-1}(\boldsymbol{b}_1 - \boldsymbol{\beta}_1)\end{pmatrix}.$$

Then using equations (6.5.6), (6.5.11), (6.5.14), and (6.5.19),

$$E(\boldsymbol{b}_2^* - \boldsymbol{\beta}_2)(\boldsymbol{b}_2^* - \boldsymbol{\beta}_2)' = \sigma^2\boldsymbol{S}_{22}^{-1} = \boldsymbol{V}(\boldsymbol{b}_2^*)\,,$$

$$E(\boldsymbol{b}_2^* - \boldsymbol{\beta}_2)(\boldsymbol{b}_1 - \boldsymbol{\beta}_1)' = \boldsymbol{0}\,,$$

and

$$E(\boldsymbol{b}_6 - \boldsymbol{b}_5)(\boldsymbol{b}_6 - \boldsymbol{b}_5)' = \begin{pmatrix}\boldsymbol{0} & \boldsymbol{0}\\ \boldsymbol{0} & \boldsymbol{V}_2 - \boldsymbol{V}_{21}\boldsymbol{V}_1^{-1}\boldsymbol{V}_{12} - \boldsymbol{V}(\boldsymbol{b}_2^*)\end{pmatrix} = \boldsymbol{V}_5 - \boldsymbol{V}_6\,.$$

This last matrix is nonnegative definite. Using (6.5.7) and (6.4.5) we therefore have the relationship

$$\tilde{\delta}(\boldsymbol{0}) = R(\hat{\boldsymbol{p}}_5) - R(\hat{\boldsymbol{p}}_6) \geqq 0\,, \tag{6.5.21}$$

where in general

$$\tilde{\delta}(\boldsymbol{V}) = R(\hat{\boldsymbol{p}}_5) - R(\hat{\boldsymbol{p}}_6) \,. \tag{6.5.22}$$

The above results may be summarized in the following theorem.

Theorem 6.4 Let there be an auxiliary estimator $\boldsymbol{b}_1^* = \boldsymbol{\beta}_1 + \boldsymbol{d}$ as in (6.5.1). Then the following relationships between the alternative predictors hold.

a) if rank $\boldsymbol{V} = J$, then

$$R(\hat{\boldsymbol{p}}_6) - R(\hat{\boldsymbol{p}}_4) \geqq 0 \tag{6.5.23}$$

as $\hat{\boldsymbol{p}}_4$ is the GLS predictor in the restricted model given by (6.2.1).

b) if $\boldsymbol{V} = \boldsymbol{0}$, then

$$R(\hat{\boldsymbol{p}}_6) \leqq R(\hat{\boldsymbol{p}}_5) \leqq R(\hat{\boldsymbol{p}}_3) \,. \tag{6.5.24}$$

c) if rank $\boldsymbol{V} < J$ and $\tilde{\delta}(\boldsymbol{0}) > 0$ where $\tilde{\delta}(\boldsymbol{V})$ is defined by (6.5.22), then in a certain neighbourhood of $\boldsymbol{V} = \boldsymbol{0}$ we have

$$R(\hat{\boldsymbol{p}}_6) < R(\hat{\boldsymbol{p}}_5) \leqq R(\hat{\boldsymbol{p}}_3) \,. \tag{6.5.25}$$

6.6 Testing linear hypotheses

The constraint $\boldsymbol{r} = \boldsymbol{R\beta}$ can also be seen as a hypothesis on the vector $\boldsymbol{\beta}$, and we can look for a significance test based upon one or other of the predictors presented in the previous section. Let us test the null hypothesis

$$H_0 : \boldsymbol{R\beta} = \boldsymbol{r} \tag{6.6.1}$$

(where $\boldsymbol{r}, \boldsymbol{R}$ are $J \times 1$, $J \times p$ respectively) against the alternative hypothesis

$$H_1 : \boldsymbol{R\beta} \neq \boldsymbol{r} \,.$$

Note that under both H_0 and H_1 the variance σ^2 has to be estimated as well as the vector $\boldsymbol{\beta}$. We shall assume normality and use the likelihood ratio approach (Anderson 1958, p. 179).

The form of the likelihood function $L(\boldsymbol{\beta}, \sigma^2)$ is as follows:

$$L(\boldsymbol{\beta}, \sigma^2) = |2\pi\sigma^2 \boldsymbol{W}|^{-\frac{1}{2}} \, exp \left\{ -\tfrac{1}{2} (\boldsymbol{y} - \boldsymbol{X\beta})' \boldsymbol{W}^{-1} (\boldsymbol{y} - \boldsymbol{X\beta} \right\} . \tag{6.6.2}$$

Using a set of Lagrangian multipliers, we find that under H_0 this is maximized when

$$\left.\begin{aligned} \boldsymbol{\beta} &= \hat{\boldsymbol{\beta}}_\omega = \boldsymbol{b}_5 \\ \text{and} \qquad & \\ \sigma^2 &= \sigma_\omega^2 = \frac{1}{n} (\boldsymbol{y} - \boldsymbol{X b}_5)' \, \boldsymbol{W}^{-1} (\boldsymbol{y} - \boldsymbol{X b}_5) \end{aligned}\right\} \tag{6.6.3}$$

where $\boldsymbol{b}_5$ is given by (6.4.3). Under H_1 the maximum likelihood estimators of $\boldsymbol{\beta}$ and σ^2 are

and

$$\left.\begin{aligned} \beta &= \hat{\beta}_\Omega = \boldsymbol{b} \\ \sigma^2 &= \sigma^2_\Omega = \frac{1}{n}(\boldsymbol{y} - \boldsymbol{Xb})' \boldsymbol{W}^{-1}(\boldsymbol{y} - \boldsymbol{Xb}) \end{aligned}\right\} \tag{6.6.4}$$

where $\boldsymbol{b}$ is the generalized least squares estimator defined in (3.4.1). (Note that $\boldsymbol{b}$ and σ^2_Ω can also be obtained by putting $\boldsymbol{R} = \boldsymbol{0}$ and $\boldsymbol{r} = \boldsymbol{0}$ in (6.4.3).)

The likelihood ratio test statistic based on the above two estimators simplifies to a ratio of quadratic forms, namely

$$\frac{(\boldsymbol{r} - \boldsymbol{Rb})' (\boldsymbol{RS}^{-1}\boldsymbol{R}')^{-1} (\boldsymbol{r} - \boldsymbol{Rb})}{(\boldsymbol{y} - \boldsymbol{Xb})' \boldsymbol{W}^{-1}(\boldsymbol{y} - \boldsymbol{Xb})}. \tag{6.6.5}$$

Under the null hypothesis, the denominator of this ratio is σ^2 times a central χ^2_J statistic (it has a noncentral distribution when the null hypothesis does not hold). The numerator of (6.6.5) is σ^2 times a central χ^2_{n-p} statistic under both H_0 and H_1. Moreover the numerator and denominator are statistically independent. Hence under H_0, (6.6.5) is $\dfrac{J}{n-p}$ times a central $F_{J,n-p}$ statistic, and critical values for testing H_0 can therefore be gleaned from the relevant table.

6.7 Summary

Section 6.2 extends the results of the previous two chapters to take account of possible auxiliary information on $\boldsymbol{\beta}$, if this information can be expressed in the form given by (6.1.1). In Section 6.3 the possibility of auxiliary information on σ^2 is used to define a family of ridge-type estimators (see (6.3.1)), whose properties are then analysed. Section 6.4 considers the gain in efficiency which can be obtained from exact prior knowledge on the value of $\boldsymbol{\beta}$.

In Section 6.5 the technique of '*piecewise*' regression is defined, and certain properties are derived. These are summarized in Theorem 6.4 (equation (6.5.23)), and generally give conditions under which particular predictors are better than others.

Finally, Section 6.6 discusses the likelihood ratio method of testing linear hypotheses.

7

Prediction and model choice

7.1 Misspecification of constraints

We have already observed how the decision to use conventional GAUSS-MARKOV estimators essentially implies complete faith in the specified model, together with the use of the usual R-criterion of optimality. In many practical situations however, some explanatory variables may be difficult or costly to record, so that the GAUSS-MARKOV estimators are less optimal, feasible, or desirable. In such situations we may instead consider estimators based on a reduced model such as

$$\boldsymbol{y} = \boldsymbol{X}_2\boldsymbol{\beta}_2 + \boldsymbol{u}\,. \tag{7.1.1}$$

Although such estimators are neither unbiased nor optimal, they can possess other desirable characteristics, which will be examined in the present section.

We consider initially the problem of predicting at a single point in time. Then the GLS predictor (see Theorem 5.2 and equation (5.2.22)) in the full model $\boldsymbol{y} = \boldsymbol{X}_1\boldsymbol{\beta}_1 + \boldsymbol{X}_2\boldsymbol{\beta}_2 + \boldsymbol{u}$ is

$$\hat{p}_3 = \boldsymbol{x}'_*\boldsymbol{b} + \boldsymbol{w}'\boldsymbol{W}^{-1}(\boldsymbol{y} - \boldsymbol{X}\boldsymbol{b})\,, \tag{7.1.2}$$

where $\boldsymbol{b}$ is the GLS estimator, $\boldsymbol{b} = \boldsymbol{S}^{-1}\boldsymbol{X}'\boldsymbol{W}^{-1}\boldsymbol{y}$. Under the standard assumptions this predictor has mean

$$\mu = \boldsymbol{x}'_*\boldsymbol{\beta} \tag{7.1.3}$$

and variance

$$\sigma_3^2 = \sigma^2(\boldsymbol{x}'_*\boldsymbol{S}^{-1}\boldsymbol{x}_* + \boldsymbol{w}'\boldsymbol{W}^{-1}\boldsymbol{w} - \boldsymbol{w}'\boldsymbol{W}^{-1}\boldsymbol{X}\boldsymbol{S}^{-1}\boldsymbol{X}'\boldsymbol{W}^{-1}\boldsymbol{w})\,. \tag{7.1.4}$$

Now adding the restrictions

$$\boldsymbol{r} = \boldsymbol{R}\boldsymbol{\beta} \tag{7.1.5}$$

leads to the GM estimator

$$\boldsymbol{b}_5 = \boldsymbol{b} + \boldsymbol{S}^{-1}\boldsymbol{R}'(\boldsymbol{R}\boldsymbol{S}^{-1}\boldsymbol{R}')^{-1}\,(\boldsymbol{r} - \boldsymbol{R}\boldsymbol{b}) \tag{7.1.6}$$

and the corresponding predictor

$$\hat{p}_5 = \boldsymbol{x}'_*\boldsymbol{b}_5 + \boldsymbol{w}'\boldsymbol{W}^{-1}(\boldsymbol{y} - \boldsymbol{X}\boldsymbol{b}_5)\,. \tag{7.1.7}$$

Note that this is identical to (7.1.2), except that $\boldsymbol{b}$ has been replaced by $\boldsymbol{b}_5$. Also note that (7.1.1) can be incorporated into (7.1.5) by writing the condition

$$[\boldsymbol{I}\ \boldsymbol{0}]\,\beta = \boldsymbol{0}\,, \tag{7.1.8}$$

where $\boldsymbol{I}$ is an identity matrix with ones corresponding to the variables not included in $\boldsymbol{X}_2$.

We may also note that

$$\hat{p}_5 = \hat{p}_3 + \hat{p}_R\,, \tag{7.1.9}$$

where

$$\begin{aligned} \hat{p}_R &= (\boldsymbol{x}_* - \boldsymbol{X}'\boldsymbol{W}^{-1}\boldsymbol{w})'\,(\boldsymbol{b}_5 - \boldsymbol{b}) \\ &= (\boldsymbol{x}_* - \boldsymbol{X}'\boldsymbol{W}^{-1}\boldsymbol{w})'\,\boldsymbol{S}^{-1}\boldsymbol{R}'(\boldsymbol{R}\boldsymbol{S}^{-1}\boldsymbol{R}')^{-1}\,(\boldsymbol{r} - \boldsymbol{R}\boldsymbol{b}) \\ &= \boldsymbol{z}'(\boldsymbol{r} - \boldsymbol{R}\boldsymbol{b})\,, \quad \text{say,} \end{aligned} \tag{7.1.10}$$

using (7.1.2), (7.1.7) and (7.1.6). If we put

$$\gamma = \boldsymbol{z}'(\boldsymbol{r} - \boldsymbol{R}\boldsymbol{\beta}) \tag{7.1.11}$$

then

$$E[\hat{p}_R] = \gamma$$

and

$$Var\,[\hat{p}_R] = \sigma^2\boldsymbol{z}'\boldsymbol{R}\boldsymbol{S}^{-1}\boldsymbol{R}'\boldsymbol{z} = \sigma_R^2, \text{ say}\,. \tag{7.1.12}$$

The covariance between $\hat{p}_3$ and $\hat{p}_R$ is

$$E[(\hat{p}_3 - \mu)\,(\hat{p}_R - \gamma)] = -\tfrac{1}{2}\,(\sigma_R^2 + g)\,, \tag{7.1.13}$$

where

$$g = 2\sigma^2\boldsymbol{w}'\boldsymbol{W}^{-1}\boldsymbol{X}\boldsymbol{S}^{-1}\boldsymbol{R}'\boldsymbol{z} + \sigma_R^2\,. \tag{7.1.14}$$

From this we deduce that $\hat{p}_5$ given by (7.1.9) has mean $(\mu + \gamma)$ and variance

$$\begin{aligned} \sigma_5^2 &= \sigma_3^2 + \sigma_R^2 - (\sigma_R^2 + g) \\ &= \sigma_3^2 - g\,. \end{aligned} \tag{7.1.15}$$

Hence the biased predictor $\hat{p}_5$ has a smaller variance than the unbiased predictor $\hat{p}_3$ when and only when g is positive.

In Section 7.3 we shall examine the conditions under which this is so, but first let us consider various criteria for comparing predictors and estimators which differ in certain respects from the criterion of R-optimality.

7.2 Criteria for comparison

We now introduce three criteria which can be used as alternatives to mean square error and R-optimality. The next section will apply these criteria to the estimators $\hat{p}_3$ and $\hat{p}_5$ discussed in Section 7.1. We start with three definitions, in all of which $\hat{\mu}_1$ and $\hat{\mu}_2$ are assumed to be estimators of μ.

Definition 7.2.1 We say that $\hat{\mu}_1$ is K_1^a-*better* than $\hat{\mu}_2$ if

$$E[(\hat{\mu}_1 - \mu)^2] < E[(\hat{\mu}_2 - \mu)^2] - a\,.$$

Equivalently we may say that $\hat{\mu}_1$ has a-smaller MSE than $\hat{\mu}_2$.

Clearly when a is zero this is just the familiar MSE principle. (In fact a may be a function of μ.) The K_1^a-criterion is transitive when $a \geqq 0$. That is, if $\hat{\mu}_1$ is K_1^a-better than $\hat{\mu}_2$, and $\hat{\mu}_2$ is K_1^a-better than $\hat{\mu}_3$ (for the same parameter μ), then $\hat{\mu}_1$ must be K_1^a-better than $\hat{\mu}_3$. (In fact $\hat{\mu}_1$ would also be K_1^{2a}-better than $\hat{\mu}_3$.)

Definition 7.2.2 We say that $\hat{\mu}_1$ is K_2^b-better than $\hat{\mu}_2$ if

$$Prob\,\{|\hat{\mu}_1 - \mu| < k\} > Prob\,\{|\hat{\mu}_2 - \mu| < k\} + b\,.$$

We also say equivalently that $\hat{\mu}_1$ is a *b-better k-neighbourhood estimator* than $\hat{\mu}_2$. This principle was used by WEBSTER (1965), and also has the property of transitivity whenever $b \geqq 0$. In a nutshell it says that the probability of $\hat{\mu}_1$ being within a particular neighbourhood of the true value exceeds the corresponding probability for $\hat{\mu}_2$ by a certain amount. If $\hat{\mu}_1$ is K_2^0-better than $\hat{\mu}_2$ for all neighbourhoods k, then the K_2^0-criterion implies the MSE criterion.

Definition 7.2.3 We say that $\hat{\mu}_1$ is K_3^c-*better* than $\hat{\mu}_2$, if

$$Prob\,\{|\hat{\mu}_1 - \mu| < |\hat{\mu}_2 - \mu|\} > \tfrac{1}{2} + c\,.$$

We also say that $\hat{\mu}_1$ lies *c-nearer* to μ than $\hat{\mu}_2$.

This principal was originated by PITMAN (1937), who would have called $\hat{\mu}_1$ a 'closer' estimator of μ than $\hat{\mu}_2$. It effectively says that $\hat{\mu}_1$ should be nearer to the true value than $\hat{\mu}_2$ more frequently than $\hat{\mu}_2$ is nearer than $\hat{\mu}_1$. The K_3-principle is not transitive, as shown by the following example based on TOUTENBURG (1973).

Suppose that X_1, X_2 and X_3 are independent random variables which take the values (3, 6, 7), (2, 5, 9), and (3, 4, 8) respectively, each with probability one third. Then if $\hat{\mu}_i = \mu + X_i$ $(i = 1, 2, 3)$ we have

$$Prob\,\{|\hat{\mu}_2 - \mu| < |\hat{\mu}_1 - \mu|\} = Prob\,\{X_2 < X_1\} = \tfrac{5}{9}\,,$$

$$Prob\,\{|\hat{\mu}_3 - \mu| < |\hat{\mu}_2 - \mu|\} = Prob\,\{X_3 < X_2\} = \tfrac{5}{9}\,,$$

and

$$Prob\,\{|\hat{\mu}_1 - \mu| < |\hat{\mu}_3 - \mu|\} = Prob\,\{X_1 < X_3\} = \tfrac{5}{9}\,.$$

Therefore $\hat{\mu}_2$ is K_3^c-better than $\hat{\mu}_1$, $\hat{\mu}_3$ is K_3^c-better than $\hat{\mu}_2$, yet $\hat{\mu}_1$ is K_3^c-better than $\hat{\mu}_3$, for any value of c up to $\frac{1}{18}$. Hence K_3-betterness does not have the transitivity property. (This example is confused by the 'tie' between the two values of 3, yet the K_3 criterion need not be transitive even in the absence of ties. It would appear however that transitivity does hold for normal variables, as well as possibly other families).

The three criteria introduced above will be applied to prediction problems in the next section, but first we wish to describe an example of criterion K_1, which is described in TOUTENBURG (1973 and 1975, p. 113).

Suppose that we have a variable X with the $N(\mu, v^2)$ distribution. For instance X may be the mean of a sample of size n from the $N(\mu, nv^2)$ distribution. $\left(\text{Note that in TOUTENBURG's notation } X \text{ here corresponds to } \bar{x}, \text{ and } v^2 \text{ correspondends to } \frac{\sigma^2}{n}\right)$. Suppose that

$$\hat{\mu}_2 = X \sim N(\mu, v^2)$$

and

$$\hat{\mu}_1 = kX \sim N(k\mu, k^2v^2)\,. \tag{7.2.4}$$

Following Definition 7.2.1, $\hat{\mu}_1$ is K_1^a-better than $\hat{\mu}_2$ whenever

$$(k-1)^2\,\mu^2 + k^2v^2 < v^2 - a\,.$$

Putting $c = \left|\frac{v}{\mu}\right|$, the coefficient of variation of X, this inequality implies that

$$(k-1)^2 + k^2c^2 < c^2 - a'\,,$$

where $a' = \frac{a}{c^2}$.

Therefore the condition for $\hat{\mu}_1$ to be K_1^a-better than $\hat{\mu}_2$ (if assuming $k^2 < 1$) is

$$c^2 > \frac{(k-1)^2 + a'}{(1-k)^2}\,. \tag{7.2.5}$$

Alternatively, this may be expressed as an inequality on k. Whenever k is in the region thus defined, $\hat{\mu}_1$ is preferable to the conventional estimator according to criterion 1.

7.3 The criteria applied to prediction

We now apply the three criteria introduced in the last section to the prediction problem outlined in Section 7.1.

a) Criterion 1

According to Definition 7.2.1, $\hat{p}_5$ given by (7.1.7) is K_1^a-better than $\hat{p}_3$, given by (7.1.2) if and only if

$$E[(\hat{p}_5 - \mu)^2] \leqq E[(\hat{p}_3 - \mu)^2] - a\,.$$

Here $\mu = \boldsymbol{x}'_* \boldsymbol{\beta} = E[y_*]$, since that is what is being predicted. Using (7.1.15) and (7.1.4), this condition is equivalent to

$$\sigma_5^2 + \gamma^2 < \sigma_3^2 - a$$

or

$$\left|\frac{\gamma}{\sigma_R}\right| < \sqrt{\frac{g - a}{\sigma_R^2}} = \delta_1, \text{ say}. \qquad (7.3.1)$$

Equation (7.3.1) gives the condition for $\hat{p}_5$ to be K_1^a-better than the usual estimator, $\hat{p}_3$. The critical parameter δ_1 is known if a is chosen as a prespecified multiple of σ^2, and if furthermore $a < g$.

b) Criterion 2

Applying the second criterion to our problem, we see that $\hat{p}_5$ is K_2^b-better than $\hat{p}_3$ if

$$Prob\,\{|\hat{p}_5 - \mu| < k\} > Prob\,\{|\hat{p}_3 - \mu| < k\} + b\,. \qquad (7.3.2)$$

The first probability in this expression can be written

$$Prob\,\{u_5 \in R_5\}\,, \qquad (7.3.3)$$

where

$$u_5 = \frac{\hat{p}_5 - \gamma - \mu}{\sigma_5}$$

and

$$R_5 = [(-k - \gamma)\,\sigma_5^{-1}, (+k - \gamma)\,\sigma_5^{-1}]\,.$$

Similarly the second probability is

$$Prob\,\{u_3 \in R_3\}\,, \qquad (7.3.4)$$

where

$$u_3 = \frac{\hat{p}_3 - \mu}{\sigma_3}$$

and

$$R_3 = [-k\sigma_3^{-1}, +k\sigma_3^{-1}]\,.$$

Note that u_5 and u_3 both have mean zero and variance one.

When we add the assumption of normality they are each $N(0, 1)$, and the probabilities given by (7.3.3) and (7.3.4) may be calculated. In general they equal $(\Phi_1 - \Phi_2)$ and $(\Phi_3 - \Phi_4)$ respectively where the Φ_i's are the relevant quantiles. With this notation (7.3.2) becomes

$$\Phi_1 - \Phi_2 > \Phi_3 - \Phi_4 + b\,.$$

This relation becomes in the prediction problem

$$\Phi\left(\frac{k - \gamma}{\sigma_5}\right) + \Phi\left(\frac{k + \gamma}{\sigma_5}\right) > 2\Phi\left(\frac{k}{\sigma_3}\right) + b\,. \qquad (7.3.5)$$

If k is given as a multiple of σ we may calculate a critical value for $\left|\frac{\gamma}{\sigma_5}\right|$, say δ^*, and from this a critical value for

$$\left|\frac{\gamma}{\sigma_R}\right|, \quad \text{say } \delta^* \sigma_5 \sigma_R^{\ 1} = \delta_2 .$$

As the lefthand of (7.3.5) is monotonically decreasing in we have the relation

$$\hat{p}_5 \text{ is } K_2^b\text{-better than } \hat{p}_3 \text{ iff } \left|\frac{\gamma}{\sigma_R}\right| < \delta_2 .$$

The critical values of δ_2 are tabulated in Figure 7.3.1.

b	$\frac{k}{\sigma}$	$\frac{\sigma_3}{\sigma} = 1$			$\frac{\sigma_3}{\sigma} = 2$		
		$\frac{\sigma_5}{\sigma} = 0.25$	0.5	0.8	0.5	1	1.5
−0.2	1	1.04	1.18	1.70	0.75	1.10	1.70
−0.2	2	1.89	1.91	2.41	1.04	1.18	1.54
−0.2	3	2.88	2.92	3.89	1.44	1.48	1.78
−0.1	1	0.98	1.03	1.35	0.63	0.90	1.30
−0.1	2	1.79	1.70	2.05	0.98	1.03	1.25
−0.1	3	2.77	2.73	3.32	1.36	1.31	1.44
0	1	0.91	0.89	0.93	0.59	0.73	0.88
0	2	1.63	1.33	1.08	0.91	1.46	0.92
0	3	2.32	1.73	1.31	1.26	1.09	1.01
0.1	1	0.83	0.71	0.21	0.53	0.57	0.30
0.2	1	0.73	0.47	*	0.46	0.39	*
0.3	1	0.49	*	*	0.39	*	*

Figure 7.3.1 Critical values of δ_2 such that, if $|\gamma/\sigma_R| < \delta_2$, then $\hat{p}_5$ is K_2^b-better than $\hat{p}_3$. (In these tables, $\sigma_R^2 = \sigma_3^2 - \sigma_5^2$, that is $w = 0$, and normality is assumed. Asterisks denote that $\hat{p}_3$ is always K_2^b-better than $\hat{p}_5$.) Source: Toutenburg (1975, p. 158).

c) Criterion 3

Using Definition 7.2.3 we have for the prediction problem the relation: $\hat{p}_5$ is K_3^c-better than $\hat{p}_3$ if

$$Prob\,\{|\hat{p}_5 - \mu| < |\hat{p}_3 - \mu|\} > \tfrac{1}{2} + c . \tag{7.3.6}$$

First we note that the stochastic event

$$|\hat{p}_5 - \mu| = |\hat{p}_3 - \mu + \hat{p}_R| < |\hat{p}_3 - \mu|$$

is equivalent to the union of the two mutually exclusive events

$$\left.\begin{array}{ll} (\hat{p}_3 - \mu) > 0 , & 0 > \hat{p}_R > -2(\hat{p}_3 - \mu) \\ \text{and} & \\ (\hat{p}_3 - \mu) < 0 , & 0 < \hat{p}_R < -2(\hat{p}_3 - \mu) . \end{array}\right\} \tag{7.3.7}$$

To evaluate the probability of (7.3.7) in the normally distributed regression model we transform $(\hat{p}_3, \hat{p}_5)$ to (p, q) where

$$p = -\frac{\hat{p}_R - \gamma}{\sigma_R} \sim N(0, 1)\,,$$

and

$$q = \frac{(\hat{p}_3 - \mu) + \frac{1}{2}(\hat{p}_R - \gamma)}{\sqrt{\sigma_3^2 - \frac{1}{4}\sigma_R^2 - \frac{1}{2}g}} \sim N(0, 1)\,.$$

Note that

$$Epq = \varrho = \frac{g}{2\sigma_R}\left(\sigma_3^2 - \tfrac{1}{4}\sigma_R^2 - \tfrac{1}{2}g\right)^{-\frac{1}{2}}.$$

The covariance ϱ is known because the factor σ cancels out. Therefore the vector (p, q) is distributed with the bivariate normal distribution

$$(p, q) \sim N_2\left[(0, 0), \begin{pmatrix} 1 & \varrho \\ \varrho & 1 \end{pmatrix}\right]. \tag{7.3.8}$$

The events (7.3.7) transform to

$$\left.\begin{aligned} &-\frac{\gamma}{\sigma_R}\frac{\varrho\sigma_R^2}{g} < q\,, \quad \frac{\gamma}{\sigma_R} < p \\ \text{and} \\ &q < -\frac{\gamma}{\sigma_R}\cdot\frac{\varrho\sigma_R^2}{g}\,, \quad p < \frac{\gamma}{\sigma_R}\,. \end{aligned}\right\} \tag{7.3.9}$$

So the probability (7.3.6) may be calculated as

$$\begin{aligned} P\{|\hat{p}_5 - \mu| < |\hat{p}_3 - \mu|\} &= 1 - P\left(\frac{\gamma}{\sigma_R} < p\right) \\ &\quad - P\left(-\frac{\gamma\varrho}{\sigma_R}\frac{\sigma_R^2}{g} < q\right) + 2P\left(-\frac{\gamma\varrho}{\sigma_R}\frac{\sigma_R^2}{g} < q\,, \frac{\gamma}{\sigma_R} < p\right) \\ &= 2 - \Phi\left(\frac{\gamma}{\sigma_R}\right) - \Phi\left(\frac{\gamma\varrho}{\sigma_R}\cdot\frac{\sigma_R^2}{g}\right) - 2P^*\left(\frac{\gamma\varrho}{\sigma_R}\frac{\sigma_R^2}{g} < q\,, \frac{\gamma}{\sigma_R} < p\right), \end{aligned} \tag{7.3.10}$$

where P^* is the c.d.f. of (7.3.8). The probability (7.3.10) is monotonically decreasing in $\left|\frac{\gamma}{\sigma_R}\right|$.

Let δ_3 denote the critical value of $\left|\frac{\gamma}{\sigma_R}\right|$ such that equality obtains in (7.3.10). Then we can assert that

$$\hat{p}_5 \text{ is } K_3^c\text{-better than } \hat{p}_3 \text{ iff } \left|\frac{\gamma}{\sigma_R}\right| < \delta_3\,. \tag{7.3.11}$$

The critical values of δ_3 are contained in Figure 7.3.2.

δ_3 \ ϱ	0.1	0.2	0.3	0.4	0.5	0.6	0.7	0.8	0.9
0	0.032	0.065	0.097	0.131	0.167	0.205	0.247	0.296	0.357
0.2	0.030	0.061	0.094	0.123	0.158	0.189	0.226	0.265	0.310
0.4	0.025	0.051	0.073	0.098	0.121	0.144	0.167	0.185	0.137
0.6	0.017	0.032	0.054	0.059	0.070	0.080	0.083	0.081	0.069
0.8	0.004	0.006	0.011	0.012	0.009	0.002	−0.011	−0.030	−0.053
1.0	−0.008	−0.017	−0.028	−0.041	−0.058	−0.078	−0.103	−0.130	−0.158
1.2	−0.022	−0.044	−0.068	−0.094	−0.123	−0.154	−0.186	−0.217	−0.245
1.4	−0.035	−0.070	−0.107	−0.145	−0.183	−0.221	−0.256	−0.288	−0.316
1.6	−0.048	−0.096	−0.144	−0.191	−0.236	−0.278	−0.314	−0.345	−0.371
1.8	−0.061	−0.119	−0.181	−0.231	−0.280	−0.325	−0.361	−0.400	−0.412
2	−0.072	−0.141	−0.210	−0.267	−0.319	−0.363	−0.397	−0.423	−0.442

Figure 7.3.2 For particular values of δ_3 and ϱ, this table contains the value of c such that

$$Prob\,\{|\hat{p}_5 - \mu| < |\hat{p}_3 - \mu|\} = \tfrac{1}{2} + c\,,$$

assuming that $w = 0$. Source: TOUTENBURG (1975, p. 159).

7.4 Deciding which predictor to use

For each of the above three criteria there is defined a unique test statistic

$$F = \left[\frac{\hat{p}_R \cdot \sigma}{\sigma_R \cdot s}\right]^2 , \tag{7.4.1}$$

where s^2 is the error sum of squares which estimates σ^2; namely

$$s^2 = (\boldsymbol{y} - \boldsymbol{X}\boldsymbol{b})' \boldsymbol{W}^{-1}(\boldsymbol{y} - \boldsymbol{X}\boldsymbol{b})\,(n - p)^{-1} . \tag{7.4.2}$$

As is well known this variable is σ^2 times a central chisquared distribution with $(n - p)$ degrees of freedom. That is

$$s^2 \sim \sigma^2(n - p)^{-1}\,\chi^2_{n-p} .$$

After some calculations it can be shown (see TOUTENBURG 1975, p. 118) that the statistic F from (7.4.1) is distributed as noncentral F. That is

$$F \sim F_{1,\,n-p}(\gamma^2\sigma_R^{-2}) , \tag{7.4.3}$$

where the noncentrality parameter is $\delta^2 = \gamma^2\sigma_R^{-2}$. This parameter provides a basis for deciding whether to use $\hat{p}_3$ or $\hat{p}_5$.

Theorem 7.1 If the unrestricted model $\boldsymbol{y} = \boldsymbol{X}_1\boldsymbol{\beta}_1 + \boldsymbol{X}_2\boldsymbol{\beta}_2 + \boldsymbol{u}$ is assumed, then we have the relation $(i = 1, 2, 3)$

$$\hat{p}_5 \text{ is } K_i\text{-}\mathit{better} \text{ than } \hat{p}_3 \text{ if } \left|\frac{\gamma}{\sigma_R}\right| < \delta_i , \tag{7.4.4}$$

where δ_1 is given in (7.3.1), δ_2 is given in Figure 7.3.1 and δ_3 is given in Figure 7.3.2 (if $w = 0$ is assumed).

Hence we may choose (7.4.4) as our null hypothesis, i.e.

$$H_0 : \delta < \delta_i \qquad (i = 1, 2, 3) .$$

and

$$H_1 : \delta \geqq \delta_i \qquad (i = 1, 2, 3) .$$

For any given Type I error level α, we may calculate a quantity $F_\alpha^{(i)}$ such that

$$P(F \geqq F_\alpha^{(i)}) = \alpha \qquad (\delta = \delta_i) .$$

Then H_0 will be accepted if and only if

$$F = \left[\frac{\hat{p}_R\sigma}{\sigma_R s}\right]^2 < F_\alpha^{(i)} \qquad (i = 1, 2, 3) .$$

If H_0 is accepted then $\hat{p}_5$ is K_i-better than $\hat{p}_3$.

7.5 Estimation and model choice

Although the K_1 criterion is easily extended to cover vector variables, this is not so for K_2 and K_3. Hence we restrict ourselves to the first criterion, and we call the vector estimator $\hat{\boldsymbol{\beta}}_1$ a MSE-better estimator than $\hat{\boldsymbol{\beta}}_2$ when

$$\Delta(\hat{\boldsymbol{\beta}}_2, \hat{\boldsymbol{\beta}}_1) = MSE\,(\hat{\boldsymbol{\beta}}_2) - MSE\,(\hat{\boldsymbol{\beta}}_1) \tag{7.5.1}$$

is non-negative definite. Applied to our problem this means that we should consider the matrix $\Delta(\boldsymbol{b}, \boldsymbol{b}_5)$. When this is non-negative definite we would tend to prefer the restricted model, upon which $\boldsymbol{b}_5$ is based. Now it may be shown that

$$\Delta(\boldsymbol{b}, \boldsymbol{b}_5) = \sigma^2 \boldsymbol{ZAZ}' ,$$

where

$$\boldsymbol{A} = \boldsymbol{RS}^{-1}\boldsymbol{R}' - \sigma^{-2}(\boldsymbol{r} - \boldsymbol{R\beta})\,(\boldsymbol{r} - \boldsymbol{R\beta})' \tag{7.5.2}$$

and

$$\boldsymbol{Z} = \boldsymbol{S}^{-1}\boldsymbol{R}'(\boldsymbol{RS}^{-1}\boldsymbol{R}')^{-1} .$$

Hence we are concerned with the conditions under which $\boldsymbol{A}$ is non-negative definite. But $\boldsymbol{A}$ has the form $\boldsymbol{A} = \boldsymbol{B} - \boldsymbol{dd}'$, so $\boldsymbol{x}'\boldsymbol{Ax}$ is positive when

$$\frac{\boldsymbol{x}'\boldsymbol{dd}'\boldsymbol{x}}{\boldsymbol{x}'\boldsymbol{Bx}} < 1 . \tag{7.5.3}$$

Therefore $\boldsymbol{x}'\boldsymbol{Ax}$ is positive for all $\boldsymbol{x}$ (i.e. $\boldsymbol{A}$ is positive definite) if the maximum value of $\boldsymbol{x}'\boldsymbol{dd}'\boldsymbol{x}/\boldsymbol{x}'\boldsymbol{Bx}$ is less than one. But this maximum value is $\boldsymbol{d}'\boldsymbol{B}^{-1}\boldsymbol{d}$. Hence the condition for $\boldsymbol{A}$ to be positive definite is $\boldsymbol{d}'\boldsymbol{B}^{-1}\boldsymbol{d} < 1$. From (7.5.2) this implies that

$$\lambda = \sigma^{-2}(\boldsymbol{r} - \boldsymbol{R\beta})'\,(\boldsymbol{RS}^{-1}\boldsymbol{R}')^{-1}\,(\boldsymbol{r} - \boldsymbol{R\beta}) < 1 . \tag{7.5.4}$$

Hence in deciding which estimator to use we should test a hypothesis on the value of λ. For further details see Toro-Vizcarrondo and Wallace (1968) and Wallace and Toro-Vizcarrondo (1969). Bibby (1972) discusses some related problems.

7.6 Summary

The first three sections of this chapter examine a certain type of mis-specified model, namely that in which certain regressor variables are deliberately omitted, and examines the two predictors defined in (7.1.2) and (7.1.7) from the points of view of the three non-standard criteria advanced in Section 7.2. It is found that under certain circumstances the prediction based upon the mis-specified model is superior to that obtained by using the correct specification. Critical values are presented in Tables 7.3.1 and 7.3.2.

In Section 7.4 an F-test is given for deciding between the two predictors, and an extension to the multivariate case is described in Section 7.5.

8

Prediction regions

8.1 Concepts and definitions

In the context of parameter estimation, methods of point estimation and interval estimation are usually developed in parallel. Similarly prediction theory contains the notion of prediction regions as a generalization of point prediction. This chapter looks at prediction regions. We concentrate our attention on interval prediction methods for multivariate normal distributions.

Let $\boldsymbol{Z} = (Z_1, \dots, Z_T)$ be a sample from $(\mathfrak{Z}, \mathfrak{U}, \boldsymbol{P}_z^\theta)$, where $\mathfrak{U}$ is the sigma-algebra consisting of the subsets of $\mathfrak{Z}$. The parameter θ is a member of the parameter space $\boldsymbol{\Omega}$, and $\boldsymbol{P}_z^\theta$ is a probability measure over $(\mathfrak{Z}, \mathfrak{U})$. Let $\{\boldsymbol{P}_z\}$ denote the class of admissible probability measures, and let $\mathfrak{Z}^T$ be the T-fold cartesian product space $Z \times Z \times \cdots \times Z$. That is $(Z_1, \dots, Z_T) \in \in \mathfrak{Z}^T$. Now we may define what is meant by a prediction region.

Definition 8.1 *A rrediction region* $\mathfrak{B}(Z_1, \dots, Z_T)$ *is a statistic* which is defined over $\mathfrak{Z}^T$ and takes value in $\mathfrak{U}$.

The prediction region $\mathfrak{B}$ maps a point $(Z_1, \dots, Z_T) \in \mathfrak{Z}^T$ into a subset of $\mathfrak{Z}$, i.e. onto a member of $\mathfrak{U}$. It is usual for the image to be a closed interval. For example let $\mathfrak{Z} = E^1$, so that $\mathfrak{Z}^T = E^T$ (T-dimensional Euclidean space). Then the closed intervals of $\mathfrak{U}$ consist of regions $[a, b]$. Hence in this case the prediction region is a statistic which defines an interval

$$\mathfrak{B}(Z_1, \dots, Z_T) = [a(Z_1, \dots, Z_T), b(Z_1, \dots, Z_T)]$$

with $a(Z_1, \dots, Z_T) \leqq b(Z_1, \dots, Z_T)$ for all points $(Z_1, \dots, Z_T) \in \mathfrak{Z}^T$.

The functions a and b may be chosen using certain criteria which will now be discussed.

Definition 8.2 *The 'coverage' of a* (fixed) *region* $\boldsymbol{A}$ *is its probability content* namely $P_Z^\theta(\boldsymbol{A})$.

Since $Z_1, \dots, Z_T$ are stochastic, clearly $\mathfrak{B}(Z_1, \dots, Z_T)$ is a stochastic region, and therefore $P_Z^\theta(\mathfrak{B})$ is a random variable. This random variable leads to two criteria for the construction of prediction regions, which will now be defined.

Definition 8.3 $\mathfrak{B}(Z_1, \dots, Z_T)$ is called a (p, q) *region*, or a *region with coverage* q *at confidence level* p, if

$$P_\theta\{P_Z^\theta(\mathfrak{B}) \geqq q\} = p \tag{8.1.1}$$

for all $\theta \in \boldsymbol{\Omega}$.

In other words, if $\mathfrak{B}$ is a (p, q) region, then the probability is p that $\mathfrak{B}$ contains at least $100q\%$ of the population.

Definition 8.4 $\mathfrak{B}(Z_1, \dots, Z_T)$ is called a q *region*, or *a region with expected coverage* q if

$$E_Z[P_Z^\theta(\mathfrak{B})] = q \tag{8.1.2}$$

for all $\theta \in \boldsymbol{\Omega}$.

This condition states that the random variable $P_Z^\theta(\mathfrak{B})$ (which lies between zero and one) should have expectation q.

Now let $\boldsymbol{Z}^* = (Z_1^*, \dots, Z_n^*)$ be a vector of future realizations of Z_i, and let $P_{\boldsymbol{Z}^*}^\theta$ be the probability measure of $\boldsymbol{Z}^*$. Then (p, q)-prediction regions and q-prediction regions may be defined as follows.

Definition 8.5 $\mathfrak{B}(Z_1, \dots, Z_T)$ is called a (p, q)-*prediction region if*

$$P_\theta\{P_{\boldsymbol{Z}^*}^\theta(\mathfrak{B}) \geqq q\} = p \tag{8.1.3}$$

for all $\theta \in \boldsymbol{\Omega}$.

Definition 8.6 $\mathfrak{B}(Z_1, \dots, Z_T)$ is called a q-*prediction region* of $\boldsymbol{Z}^*$ if

$$E_Z[P_{\boldsymbol{Z}^*}^\theta(\mathfrak{B})] = q \tag{8.1.4}$$

for all $\theta \in \boldsymbol{\Omega}$.

In the special case where $\boldsymbol{Z}^*$ has the same distribution as $\boldsymbol{Z}$, these two definitions are equivalent to (8.1.1) and (8.1.2).

8.2 On q-prediction intervals

Suppose now that we have a sample statistic $z = z(Z_1, \dots, Z_T)$ with probability density $p_T(z \mid \theta)$ and corresponding probability measure $P(. \mid \theta)$. We want to construct an interval

$$\mathfrak{B}(z_1, \dots, z_T) = \delta(z) = [a(z), b(z)]$$

which contains a stochastic variable z^*, where z^* is the stochastic realization of the variable z at time T^*. We assume that z^* has the density $p_{T^*}(z^* \mid \theta)$ $[z^* \in \boldsymbol{Z}^*]$ and the corresponding probability measure $P_{T^*}(. \mid \theta)$. The common parameter θ in both densities ensures the possibility of predicting z^* on the basis of z.

We assume that z and z^* are independent for given θ. Let $P_{TT^*}(. \mid \theta)$ be the joint probability measure for z and z^*. Then (8.1.4) suggests the construction of a prediction interval $\delta(z)$ according to the condition

$$P_{TT^*}\{(z, z^*) : z^* \in \delta(z) \mid \theta\} = E_Z\{P_{T^*}(\delta(z) \mid \theta\} = q \tag{8.2.1}$$

for all $\theta \in \boldsymbol{\Omega}$.

We confine ourselves to normal samples, i.e. we assume

$$\left.\begin{array}{l} \boldsymbol{Z} = E^T\,, \qquad \boldsymbol{\theta} = (\mu, \sigma^2) \in \boldsymbol{\Omega} = \{-\infty < \mu < \infty, \sigma^2 > 0\}\,, \\ z_t \sim N(\mu, \sigma^2)\,, \qquad z_t \text{ and } z_{t'} \text{ independent } t \neq t',\ z^* \sim N(\mu, \sigma^2)\,. \end{array}\right\} \qquad (8.2.2)$$

In constructing an interval $\delta(z) = \big(a(z), b(z)\big)$ we choose the statistic $z = (m, s)$ which is a sufficient statistic for $\theta = (\mu, \sigma^2)$. Suppose that $T^{-1} = h$ and $T - 1 = v$. Then we have

$$m = T^{-1}\Sigma z_t \sim N(\mu, \sigma^2 h)$$

and

$$s^2 = v^{-1}\Sigma(z_t - m)^2 \sim v^{-1}\sigma^2\chi_v^2\,. \qquad (8.2.3)$$

In this case the elements of z (m and s) are independently distributed, although this is not true in general.

We confine ourselves to intervals of the type

$$\delta(z) = [a(z), b(z)] = [m - k_1 s, m + k_2 s] \qquad (8.2.4)$$

These intervals feature widely in the theory of confidence interval estimation. Another justification for this choice is that (8.2.4) is invariant under a general class of linear transformations of the sample variables (Aitchison and Dunsmore (1968), Guttman (1970)).

Relative to (8.2.1) the values k_1, k_2 of $\delta(z)$ have to be determined such that

$$q = \int_{-\infty}^{\infty}\int_{0}^{\infty} P_{T^*}\left\{\frac{m-\mu}{\sigma} - k_1\frac{s}{\sigma} \leqq \frac{Z^* - \mu}{\sigma} \leqq \frac{m-\mu}{\sigma} + k_2\frac{s}{\sigma}\right\} \mathrm{d}p_1(m)\,\mathrm{d}p_2(s) \qquad (8.2.5)$$

holds. We transform m and s by

$$M = \frac{m - \mu}{\sigma\sqrt{h}} \sim N(0, 1)\,, \qquad S = \frac{s}{\sigma} \sim \sqrt{\chi_v^2 v^{-1}}\,.$$

Based on this, on the independence of z and z^* and on the relations

$$p_1(m)\,\mathrm{d}m = \tilde{p}_1(M)\,\mathrm{d}M$$

$$p_2(s)\,\mathrm{d}s = \tilde{p}_2(S)\,\mathrm{d}S$$

(where $\tilde{p}_1(M)$ is the $N(0, 1)$-density, $\tilde{p}_2(S)$ is the $\sqrt{\chi_v^2 v^{-1}}$-density) we get the following equation for the calculation of k_1, k_2 which is equivalent to (8.2.5)

$$q = \int_{-\infty}^{\infty}\int_{0}^{\infty} [\Phi(M\sqrt{h} + k_2 S) - \Phi(M\sqrt{h} - k_1 S)]\,\mathrm{d}\tilde{p}_1(M)\,\mathrm{d}\tilde{p}_2(S)\,. \qquad (8.2.6)$$

Let $u \sim N(0, 1)$ be a variable which is independent of M and S.

Then the following transformation can be made:

$$\left.\begin{aligned} \Phi(M\sqrt{h}+k_2S) &= P\left(\frac{u-M\sqrt{h}}{S}\leqq k_2\right)=P\left(t_v\leqq\frac{k_2}{\sqrt{1+h}}\right),\\ \Phi(M\sqrt{h}-k_1S) &= P\left(\frac{u+M\sqrt{h}}{S}\geqq k_1\right)=P\left(t_v\geqq\frac{k_1}{\sqrt{1+h}}\right).\end{aligned}\right\}\tag{8.2.7}$$

Thus we get the equation determining k_1 and k_2

$$q=P\left(t_v\leqq\frac{k_2}{\sqrt{1+h}}\right)-P\left(t_v\geqq\frac{k_1}{\sqrt{1+h}}\right).\tag{8.2.8}$$

There are two standard ways of determining k_1 and k_2.

Case (a). Two-sided symmetric interval

We choose $k_1 = k_2 = k^*$ and derive from (8.2.7)

$$q = 2P\left(t_v \leqq \frac{k^*}{\sqrt{1+h}}\right) - 1$$

i.e. $\delta(z) = (m - k^*s, m + k^*s)$ with

$$k^* = \sqrt{1+h}\, t_{v,(1-q)/2}\,.\tag{8.2.9}$$

Here $t_{v,\alpha}$ is the $(1-\alpha)$-quantile of Student's t with v degrees of freedom: $P(t \geqq t_{v,\alpha}) = \alpha$. If the interval is finite we are interested not only in the given coverage q but also in the expected length, namely

$$l(\delta) = 2\sqrt{1+h}\, t_{v,\frac{1}{2}(1-q)} E[s]\,.\tag{8.2.10}$$

If no other criterion of goodness is given, the statistician will generally prefer the shorter of two given intervals i.e. δ_1 will be preferred to δ_2 if $l(\delta_1) < l(\delta_2)$.

Case (b). One-sided infinite interval

The infinite interval open to the left is obtained by choosing $k_1 = \infty$, $k_2 = k_2^*$. This gives

$$\delta_2(z) = [-\infty, m + k_2^* s]$$

with

$$k_2^* = \sqrt{1+h}\, t_{v,1-q}\,.\tag{8.2.11}$$

The infinite interval open to the right is obtained similarly.

8.3 On q-intervals in regression analysis

Suppose once again that we have the standard normal regression model, namely

$$\left.\begin{aligned} \boldsymbol{y} &= \boldsymbol{X\beta}+\boldsymbol{u}\sim N(\boldsymbol{X\beta},\sigma^2\boldsymbol{W})\,,\\ y_* &= \boldsymbol{x}_*'\boldsymbol{\beta}+u_*\sim N(\boldsymbol{x}_*'\boldsymbol{\beta},\sigma^2 w_*)\,,\\ E[\boldsymbol{u}u_*] &= \boldsymbol{0}\,.\end{aligned}\right\}\tag{8.3.1}$$

In terms of the general formulation presented on p. 120, we get

$$y_* = \boldsymbol{Z}^* , \quad \boldsymbol{\theta} = (\beta, \sigma^2) , \quad \text{and} \quad z = (\boldsymbol{x}'_* \hat{\boldsymbol{\beta}}, sw_*) \tag{8.3.2}$$

where s is given by (7.4.2). Now clearly,

$$\boldsymbol{x}'_* \hat{\boldsymbol{\beta}} \sim N(\boldsymbol{x}'_* \boldsymbol{\beta}, \boldsymbol{x}'_* \boldsymbol{V}_{\hat{\beta}} \boldsymbol{x}_*)$$

since $\hat{\boldsymbol{\beta}}$ is an unbiased estimator of $\boldsymbol{\beta}$. Therefore, using a transformation similar to (8.2.7) but putting $u \sim N(0, w_*^2)$, and

$$h = \sigma^{-2} \boldsymbol{x}'_* \boldsymbol{V}_{\hat{\beta}} \boldsymbol{x}_*$$

we can derive various types of q-intervals for y_*, which are given in the following theorems.

Theorem 8.1 In the normal regression model (8.3.1), the two-sided symmetric q-intervals for y_* have the following form

$$\delta(\boldsymbol{x}'_* \hat{\boldsymbol{\beta}}, sw_*) = (\boldsymbol{x}'_* \hat{\boldsymbol{\beta}} - k^* sw_*, \boldsymbol{x}'_* \hat{\boldsymbol{\beta}} + k^* sw_*) \tag{8.3.3}$$

where

$$k^* = \sqrt{w_*^2 + \sigma^{-2} \boldsymbol{x}'_* \boldsymbol{V}_{\hat{\beta}} \boldsymbol{x}_*} \; t_{T-p, (1-q)/2}$$

and $\hat{\boldsymbol{\beta}}$ is an unbiased estimator of $\boldsymbol{\beta}$. This has expected length

$$l(\delta) = 2E(s) \, t_{T-p,(1-q)/2} \sqrt{w_*^2 + \sigma^{-2} \boldsymbol{x}'_* \boldsymbol{V}_{\hat{\beta}} \boldsymbol{x}_*} \, . \tag{8.3.4}$$

The one-sided infinite intervals are

$$\delta_2(\boldsymbol{x}'_* \hat{\boldsymbol{\beta}}, sw_*) = (-\infty, \boldsymbol{x}'_* \hat{\boldsymbol{\beta}} + k_2^* sw_*)$$

and

$$\delta_1(\boldsymbol{x}'_* \hat{\boldsymbol{\beta}}, sw_*) = (\boldsymbol{x}'_* \hat{\boldsymbol{\beta}} - k_1^* sw_*, \infty)$$

respectively, where

$$k_1^* = k_2^* = \sqrt{w_*^2 + \sigma^{-2} \boldsymbol{x}'_* \boldsymbol{V}_{\hat{\beta}} \boldsymbol{x}_*} \; t_{T-p, 1-q} \, .$$

Taking into account the fact that the risk function is $R(\hat{\boldsymbol{\beta}}) = tr\, \boldsymbol{A} \boldsymbol{V}_{\hat{\beta}}$ whenever $\hat{\boldsymbol{\beta}}$ is unbiased, we have the following result.

Theorem 8.2 The R-optimal unbiased estimators yield q-prediction intervals with minimal length.

From this theorem it follows that the interval

$$\delta(\boldsymbol{x}'_* \boldsymbol{b}, sw_*) = [\boldsymbol{x}'_* \boldsymbol{b} - k^* sw_*, \boldsymbol{x}'_* \boldsymbol{b} + k^* sw_*]$$

is optimal so long as there is no prior information on $\boldsymbol{\beta}$. On the other hand, if the statistician has restrictions or prior information on $\boldsymbol{\beta}$ such as that given in equations (6.1.1) and (6.1.2), then the following results may be obtained

$$l(\hat{\beta}_4) \leqq l(\boldsymbol{b}_5) \leqq l(\boldsymbol{b}) \quad \text{if } \boldsymbol{V} \text{ has full rank}, \tag{8.3.5}$$

$$l(\boldsymbol{b}_6 \mid \boldsymbol{V} = \boldsymbol{0}) \leqq l(\boldsymbol{b}_5) \leqq l(\boldsymbol{b}) \quad \text{if} \quad \boldsymbol{V} = \boldsymbol{0}, \tag{8.3.6}$$

$$l(\boldsymbol{b}_6) < l(\boldsymbol{b}_5) \leqq l(\boldsymbol{b}) \quad \text{in a neighbourhood of } \boldsymbol{V} = \boldsymbol{0}. \tag{8.3.7}$$

(The estimators used in these results are defined in (3.4.1), (6.2.8), (6.4.4) and (6.5.7).)

The above concepts and results may be generalized to the case where an $n \times 1$ vector $\boldsymbol{y}_*$ must be predicted. This gives the concept of "*q-covering ellipsoids*". We assume that

$$\boldsymbol{y}_* \sim N(\boldsymbol{X}_*\boldsymbol{\beta}, \sigma^2\boldsymbol{W}_*) \quad \text{and} \quad \boldsymbol{y} \sim N(\boldsymbol{X}\boldsymbol{\beta}, \sigma^2\boldsymbol{W}). \tag{8.3.8}$$

We confine ourselves again to unbiased estimators $\hat{\boldsymbol{\beta}}$ of $\boldsymbol{\beta}$ and to s^2 from (7.4.2) as estimator of σ^2. Then we have

$$Q(\hat{\boldsymbol{\beta}}) = n^{-1}(\boldsymbol{y}_* - X_*\hat{\boldsymbol{\beta}})' (\sigma^{-2}\boldsymbol{X}_*\boldsymbol{V}_{\hat{\beta}}\boldsymbol{X}_*' + \boldsymbol{W}_*)^{-1} (\boldsymbol{y}_* - \boldsymbol{X}_*\hat{\boldsymbol{\beta}}) \sim \sigma^2 n^{-1}\chi_n^2 .$$

The following statistic is chosen

$$z(\hat{\boldsymbol{\beta}}) = Q(\hat{\boldsymbol{\beta}})\, s^{-2} . \tag{8.3.9}$$

According to the prerequisites already advanced we have

$$z(\boldsymbol{b}) \sim F_{n,\, n-p} \cdot$$

from which the q-prediction ellipsoid for $\boldsymbol{y}_*$ (8.3.8) in the unrestricted model $\boldsymbol{y} = \boldsymbol{X\beta} + \boldsymbol{u}$ follows:

$$\mathfrak{B}[z(\boldsymbol{b})] = \{\boldsymbol{y}_* : (\boldsymbol{y}_* - \boldsymbol{X}_*\boldsymbol{b})' (\boldsymbol{X}_*\boldsymbol{S}^{-1}\boldsymbol{X}_*' + \boldsymbol{W}_*)^{-1} \times$$
$$\times (\boldsymbol{y}_* - \boldsymbol{X}_*\boldsymbol{b}) < s^2 n F_{n,\, n-p,\, 1-q}\} \tag{8.3.10}$$

So $\mathfrak{B}[z(\boldsymbol{b})]$ consists of all inner points of the ellipsoid

$$(\boldsymbol{y}_* - \boldsymbol{X}_*\boldsymbol{b})' ([\boldsymbol{X}_*\boldsymbol{S}^{-1}\boldsymbol{X}_*' + \boldsymbol{W}_*)\, s^2 n F_{n,\, n-p,\, 1-q}]^{-1} (\boldsymbol{y}_* - \boldsymbol{X}_*\boldsymbol{b}) = 1 \tag{8.3.11}$$

which in centered around $\boldsymbol{X}_*\boldsymbol{b}$.

As in the case $n = 1$ (only a single regressand y_* to be predicted) we should now investigate the influence of prior information on the volume of the prediction ellipsoid (see TOUTENBURG (1975) for multivariate models). The length of the interval corresponds to the volume of the ellipsoid. Let $\boldsymbol{V}_n$ be the volume of the n-dimensional unit sphere. Then the volume of an arbitrary ellipsoid $\boldsymbol{a}'\boldsymbol{A}\boldsymbol{a} = 1$ ($\boldsymbol{A}$ positive definite) is

$$\boldsymbol{V}_{\boldsymbol{A}} = \boldsymbol{V}_n\, |\boldsymbol{A}|^{-1/2} .$$

Based on this we get the expected volume of $\mathfrak{B}[z(\boldsymbol{b})]$ as

$$\boldsymbol{V}(\boldsymbol{b}) = \boldsymbol{V}_n \sqrt{nF_{n,\, n-p,\, 1-q}}\, E(s)\, [\boldsymbol{X}_*\boldsymbol{S}^{-1}\boldsymbol{X}_*' + \boldsymbol{W}_*]^{1/2} . \tag{8.3.12}$$

If $\hat{\boldsymbol{\beta}}$ is chosen such that $Q(\hat{\boldsymbol{\beta}})$ and s^2 are distributed independently we have

$$\mathfrak{B}[z(\hat{\boldsymbol{\beta}})] = \{y_* : z(\hat{\boldsymbol{\beta}}) < F_{n, n-p, 1-q}\} , \tag{8.3.13}$$

$$V(\hat{\boldsymbol{\beta}}) = V_n \sqrt{nF_{n, n-p, 1-q}}\, E(s)\, [\sigma^{-2}\mathbf{X}_* \boldsymbol{V}_{\hat{\beta}} \mathbf{X}_*' + \boldsymbol{W}_*]^{1/2} \tag{8.3.14}$$

Therefore the volume depends on $\boldsymbol{V}_{\hat{\beta}}$ and we get the following result.

Theorem 8.3 In any of the models of normal regression considered above, the corresponding unbiased R-optimal estimator $\hat{\boldsymbol{\beta}}$ of $\boldsymbol{\beta}$ yields the q-predition region $\mathfrak{B}[z(\hat{\boldsymbol{\beta}})]$ with minimal volume.

8.4 On (*p*, *q*)-prediction intervals

Under general assumptions on the probability measure P_Z^θ the *coverage* of the region $\mathfrak{B}$ possesses a highly complicated distribution. For this reason we confine ourselves to normally distributed samples, i.e. to the model (8.3.1) of normal regression. We choose the statistic $z = (\boldsymbol{x}_*'\boldsymbol{b}, sw_*)$ from (8.3.2) and construct an interval for y_* which is symmetrical around $\boldsymbol{x}_*'\boldsymbol{b}$, namely

$$\delta(\boldsymbol{x}_*'\boldsymbol{b}, sw_*) = (\boldsymbol{x}_*'\boldsymbol{b} - ksw_*, \boldsymbol{x}_*'\boldsymbol{b} + ksw_*) . \tag{8.4.1}$$

Here $k = k(q, p)$ has to be determined according to Definition 8.5 such, that

$$P\{P[y_* \in \delta(\boldsymbol{x}_*'\boldsymbol{b}, sw_*) \geqq q]\} = p \tag{8.4.2}$$

holds for all $\boldsymbol{\theta} = (\beta, \sigma^2)$.

Let $\tilde{\boldsymbol{u}} = u_{(1-q)/2}$ represent the $(1 + q)/2$-quantile of $N(0, 1)$-distribution. u_α is defined by $P(u \geqq u_\alpha) = \alpha$. The value of $\tilde{u}$ can be found from tables by interpolation, e.g. $q = 0.8$ gives the value $\tilde{u} = 1.284$. When $y_* \sim N(x_*'\boldsymbol{\beta}, \sigma^2 w_*^2)$ as in (8.3.1) we get

$$P(\boldsymbol{x}_*'\boldsymbol{\beta} - \tilde{u}\sigma w_* < y_* < \boldsymbol{x}_*'\boldsymbol{\beta} + \tilde{u}\sigma w_*) = q , \tag{8.4.3}$$

and from this the q-interval

$$\delta^* = (\boldsymbol{x}_*'\boldsymbol{\beta} - \tilde{u}\sigma w_*, x_*'\boldsymbol{\beta} + \tilde{u}\sigma w_*) . \tag{8.4.4}$$

This contains unknown parameters. To fulfill condition (8.4.2) we construct a confidence interval for δ^* at confidence level p of the form

$$\delta(\boldsymbol{x}_*'\boldsymbol{b}, sw_*) = (\boldsymbol{x}_*'\boldsymbol{b} - ks\sqrt{\boldsymbol{x}_*'\mathbf{S}^{-1}\boldsymbol{x}_* + w_*^2}, \boldsymbol{x}_*'\boldsymbol{b} + ks\sqrt{\boldsymbol{x}_*'\mathbf{S}^{-1}\boldsymbol{x}_* + w_*^2}) \tag{8.4.5}$$

i.e. we calculate $k = k(q, p)$ from the equation

$$P(|\boldsymbol{x}_*'(\boldsymbol{\beta} - \boldsymbol{b}) \pm \tilde{u}\sigma w_*| \leqq ks\sqrt{\boldsymbol{x}_*'\mathbf{S}^{-1}\boldsymbol{x}_* + w_*^2}) = p . \tag{8.4.6}$$

To prove Theorem 8.4 we need the following lemma.

Lemma (LIEBERMANN and MILLER (1963)) Let a_i, b_i be arbitrary real numbers ($i = 1, \dots, N$) with $a_1 b_1 \neq 0$, $A > 0$. Then $\sum a_i^2 \leqq A^2$ iff

$$|\sum a_i b_i| \leq A(\sum b_i^2)^{1/2} \qquad \text{holds for all } b_1, \dots, b_N .$$

With the help of this lemma we prove

Theorem 8.4 In model (8.3.1) the following equation holds

$$\begin{aligned} &P\{|\boldsymbol{x}_*'(\boldsymbol{\beta} - \boldsymbol{b}) \pm u^*\sigma w_*| \leqq ks \sqrt{\boldsymbol{x}_*'\boldsymbol{S}^{-1}\boldsymbol{x}_* + w_*^2}\} \\ &\quad = P\{u_1^2 + \cdots + u_K^2 + u^{*2} \leqq (n-p)^{-1} k^2 \chi_{n-p}^2\} \end{aligned}$$

where the u_i are i.i.d. $N(0, 1)$ and independent of the χ^2-distributed variable.

Proof. Since $\boldsymbol{S}^{-1}$ is positive definite there exists a nonsingular matrix $\boldsymbol{Q}$ such that $\boldsymbol{Q}'\boldsymbol{S}^{-1}\boldsymbol{Q} = \boldsymbol{I}$. Let

$$\tilde{\boldsymbol{\beta}} = \boldsymbol{Q}\boldsymbol{\beta}, \quad \tilde{\boldsymbol{b}} = \boldsymbol{Q}\boldsymbol{b}, \quad \tilde{\boldsymbol{x}}_* = (\boldsymbol{Q}^{-1})' \boldsymbol{x}_* .$$

Then we get

$$\boldsymbol{x}_*'\boldsymbol{S}^{-1}\boldsymbol{x}_* = \tilde{\boldsymbol{x}}_*'\tilde{\boldsymbol{x}}_* = \sum_{i=1}^{p} \tilde{x}_{i*}^2, \qquad \tilde{\boldsymbol{b}} \sim N(\tilde{\boldsymbol{\beta}}, \sigma^2\boldsymbol{I}),$$

$$\boldsymbol{x}_*'(\boldsymbol{\beta} - \boldsymbol{b}) = \tilde{\boldsymbol{x}}_*'(\tilde{\boldsymbol{\beta}} - \tilde{\boldsymbol{b}}), \quad \tilde{\boldsymbol{b}} \text{ independent of } s^2,$$

and with the help of the above lemma

$$\begin{aligned} &P\{|\boldsymbol{x}_*'(\boldsymbol{\beta} - \boldsymbol{b}) \pm \tilde{u}\sigma w_*| \leqq ks \sqrt{\boldsymbol{x}_*'\boldsymbol{S}^{-1}\boldsymbol{x}_* + w_*^2}\} \\ &\qquad = P\{|\tilde{\boldsymbol{x}}_*'(\tilde{\boldsymbol{\beta}} - \tilde{\boldsymbol{b}}) \pm \tilde{u}\sigma w_*| \leqq ks \sqrt{\tilde{\boldsymbol{x}}_*'\tilde{\boldsymbol{x}}_* + w_*^2}\} \\ &\qquad = P\{\sum u_i^2 + \tilde{u}^2 \leqq k^2(n-p)^{-1} \chi_{n-p}^2\} . \end{aligned}$$

Then the stochastic variable

$$(n-p)(\sum u_i^2 + \tilde{u}^2)/(p+1)\chi_{n-p}^2 \tag{8.4.7}$$

is distributed as $F_{p+1, n-p}(\tilde{u}^2)$ with noncentrality parameter $\tilde{u}^2$. This way the required constant $k = k(q, p)$ is found following (8.4.5) as the solution to

$$k^2(p+1)^{-1} = F_{p+1, n-p, 1-p}(u_{(1-q)/2}^2) = F^* \tag{8.4.8}$$

where F^* is the p-quantile of the noncentral $F_{p+1, n-p}(u_{(1-q)/2}^2)$ and $u_{(1-q)/2}$ is the $(1+q)/2$-quantile of $N(0, 1)$.

If the statistic z is chosen more generally as $z = (\boldsymbol{x}_*'\hat{\boldsymbol{\beta}}, sw_*)$, where $\hat{\boldsymbol{\beta}}$ is an unbiased estimator of $\boldsymbol{\beta}$ such that $\sum u_i^2$ in (8.4.7) is independent of s^2, then the following result holds.

Theorem 8.5 Under model (8.3.1) and the statistic $z = (\boldsymbol{x}_*'\hat{\boldsymbol{\beta}}, sw_*)$ with s^2 from (7.4.2), the (p, q)-prediction interval based on z is

$$\delta(\boldsymbol{x}_*'\hat{\boldsymbol{\beta}}, sw_*) = \left(\boldsymbol{x}_*'\hat{\boldsymbol{\beta}} - sa_{\hat{\beta}}\sqrt{(p+1)F^*}, \boldsymbol{x}_*'\hat{\boldsymbol{\beta}} + sa_{\hat{\beta}}\sqrt{(p+1)F^*}\right) \tag{8.4.9}$$

where $a_{\hat{\beta}}^2 = \sigma^{-2}\boldsymbol{x}_*'\boldsymbol{V}_{\hat{\beta}}\boldsymbol{x}_* + w_*^2$ and F^* is given by (8.4.8).
The expected length is

$$l(\delta) = 2E(s)\sqrt{(p+1)F^*}\sqrt{\sigma^{-2}\boldsymbol{x}_*'\boldsymbol{V}_{\hat{\beta}}\boldsymbol{x}_* + w_*^2} . \tag{8.4.10}$$

We see that the R-optimal estimates $\hat{\boldsymbol{\beta}}$ of $\boldsymbol{\beta}$ in the restricted model yield (p, q)-prediction regions which are optimal in the sense of possessing minimal expected length $J(\delta)$. (See GUTTMAN (1970) for further investigations, e.g. for the approximation and calculation of k in tables, the derivation of (p, q)-ellipsoids under utility functions, etc.)

8.5 Linear utility functions

8.5.1 Introduction. As we remember, the q-prediction intervals for z^* had to be determined according to (8.2.1) in such a way that the intervals $\delta(z_1), \delta(z_2), \ldots$ based on the independent samples $z_1, z_2, \ldots$ contain z^* in q per cent of the samples. The statistician's influence on the expected length of the intervals was ensured only by the choice of the statistic z, especially by the choice of the R-optimal estimates. In practice there is often a prior preference for some regions of $\mathfrak{U}$. This preference can be handled mathematically by using a utility function of $\delta(z)$ in the following manner.

Let $V(\delta, z^*)$ be a given utility function of $\delta(z)$. $V(\delta, z^*)$ expresses the utility of $\delta(z)$ in a qualitative or quantitative manner if z^* is realized. Since z and z^* are stochastic, the same is true for $V(\delta, z^*)$, which has the conditional expectation (expected utility of $\delta(z)$)

$$\overline{V}(\delta, \theta) = \int_{Z^*} V(\delta, z^*)\, p_{T^*}(z^* \mid \theta)\, dz^* = E_{Z^*}(V(\delta, z^*) \mid z)\,. \quad (8.5.1)$$

We shall call $\overline{V}(\delta, \theta)$ the advantage of $\delta(z)$.

Definition 8.7 A prediction interval $\delta^*(z)$ from a class of admissible intervals $\{\delta(z)\}$ is called *V-optimal* if the following relationship holds:

$$\max_{\{\delta(z)\}} \int_Z \overline{V}(\delta, \theta)\, p_T(z \mid \theta)\, dz = \int_Z \overline{V}(\delta^*, \theta)\, p_T(z \mid \theta)\, dz\,. \quad (8.5.2)$$

Let $Z^* = E^1$ (1-dimensional EUCLIDEAN space). Then the prediction regions for z^* are subregions of E^1, i.e. intervals

$$\delta(z) = (r_1(z), r_2(z)) \quad \text{with} \quad r_1(z) \leqq r_2(z)$$

for all $z \in Z$. Let the utility function have the general form

$$V(\delta, z^*) = \begin{cases} h_1(z^*, \delta) & \text{for } z^* \in \delta(z) \\ h_2(z^*, \delta) & \text{for } z^* \notin \delta(z) \end{cases} \quad (8.5.3)$$

i.e. in the particular case where $\delta(z) = (r_1, r_2)$,

$$V(r_1, r_2, z^*) = \left\{\begin{array}{ll} h_{21}(z^*, r_1) & (z^* \leqq r_1) \\ h_1(z^*, r_1, r_2) & (r_1 < z^* < r_2) \\ h_{22}(z^*, r_2) & (r_2 \leqq z^*)\,. \end{array}\right\} \quad (8.5.4)$$

We assume that

$h_{21}(z^*, r_1)$ is monotonically decreasing in $(r_1 - z^*)$.

$h_{22}(z^*, r_2)$ is monotonically decreasing in $(z^* - r_2)$.

In practice utility functions which are linear in $|z^* - r_i|$ have a special interest. For instance the function

$$V(r_1, r_2, z) = \begin{cases} \lambda(z^* - r_1) & [z^* \leqq r_1] \\ r_1 - z^* & [r_1 < z^* \leqq \frac{1}{2}(r_1 + r_2)] \\ z^* - r_2 & [\frac{1}{2}(r_1 + r_2) < z^* < r_2 \\ \lambda(r_2 - z^*) & [z^* \geqq r_2] . \end{cases} \tag{8.5.5}$$

Here λ is a so-called "*relative cost factor*". Let v_1 be the loss in observing $z^* \notin (r_1, r_2)$ and v_2 the loss in observing $z^* \in (r_1, r_2)$. If the statistician values loss v_1 higher than v_2 he will choose $\lambda > 1$.

Depending on the given utility function $V(\delta, z^*)$, an arbitrary subregion of $\boldsymbol{Z}^*$ may result as the V-optimal prediction region. In the case $\delta^*(z) = \boldsymbol{Z}^*$ we get the trivial prediction "$z^* \in \boldsymbol{Z}^*$" which is always true and therefore useless and to be avoided. For some types of utility functions which do not depend explicitly on the region's volume, it is possible to notice before optimization whether $\delta^*(z)$ is a proper subregion of $\boldsymbol{Z}^*$. Up to now only necessary conditions on V are known for ensuring the finiteness of (r_1, r_2) (see TOUTENBURG, 1971).

One very simple utility function yielding $\delta^*(z) = \boldsymbol{Z}^*$ is the so-called characteristic or indicator function of $\delta(z)$

$$V_0(\delta, z^*) = \begin{cases} 1 & \textit{if } z^* \in \delta(z) \\ 0 & \textit{otherwise} . \end{cases} \tag{8.5.6}$$

As $V_0(\delta, z^*)$ does not depend on the region's volume we get

$$\overline{V}(\delta, \theta) = P_{T^*}(\delta(z) \mid \theta)$$

and

$$\int_Z \overline{V}(\delta, \theta)\, p_T(z \mid \theta)\, \mathrm{d}z = P_{TT^*}\{(z, z^*): z^* \in \delta(z)\} .$$

The maximum 1 is reached for $\delta^*(z) = \boldsymbol{Z}^*$.

8.5.2 Normally distributed populations — two-sided symmetric intervals. We assume that (8.2.2) holds and choose the following type of intervals

$$\delta(z) = \big(r_1(z), r_2(z)\big) = (m - ks, m + ks) .$$

With $z = (m, s)$ the function $V(r_1, r_2, z)$ from (8.5.6) is then

$$V(k, z^*) = \begin{cases} \lambda(z^* - m + ks) & [z^* \leqq m - ks] \\ m - ks - z^* & [m - ks < z^* \leqq m] \\ z^* - m - ks & [m < z^* < m + ks] \\ \lambda(m + ks - z^*) & [z^* \geqq m + ks] , \end{cases} \tag{8.5.7}$$

so that

$$\bar{V}(k, \theta) = \lambda \int_{-\infty}^{m-ks} (z^* - m + ks)\, p_{T*}(z^* \mid \theta)\, dz^*$$
$$+ \int_{m-ks}^{m} (m - ks - z^*)\, p_{T*}(z^* \mid \theta)\, dz^*$$
$$+ \int_{m}^{m+ks} (z^* - m - ks)\, p_{T*}(z^* \mid \theta)\, dz^*$$
$$+ \lambda \int_{m+ks}^{\infty} (m + ks - z^*)\, p_{T*}(z^* \mid \theta)\, dz^* .$$

We have

$$\frac{\partial \bar{V}(k, \theta)}{\partial k} = s[\lambda - (1 + \lambda)\, P_{T*}(m - ks < z^* < m + ks \mid \theta)] .$$

Therefore the equation determining the optimal $k = k^*$ is of the form

$$\frac{\partial}{\partial k} \int_Z \bar{V}(k, \theta)\, p_T(z \mid \theta)\, dz$$
$$= \int_{-\infty}^{\infty} \int_0^{\infty} [\lambda - (1 + \lambda)\, P_{T*}(m - ks, m + ks \mid \theta)]\, s p_1(m)\, p_2(s)\, dm\, ds = 0 . \tag{8.5.8}$$

After transforming according to

$$M = \frac{m - \mu}{\sigma \sqrt{h}} , \qquad S = \frac{s \sqrt{v}}{\sigma \sqrt{v + 1}} ,$$

the relation (8.5.8) becomes (apart from a constant term)

$$0 = \int_{-\infty}^{\infty} \int_0^{\infty} \left[\lambda - (1 + \lambda) \left\{\Phi\left(M \sqrt{h} + kS \sqrt{\frac{v+1}{v}}\right) - \Phi\left(M \sqrt{h} - kS \sqrt{\frac{v+1}{v}}\right)\right\}\right] \tilde{p}_1(M)\, \tilde{p}_2^*(S)\, dM\, dS .$$

Here $\tilde{p}_1(M)$ is the $N(0, 1)$-density and $\tilde{p}_2^*(S)$ the $[\chi^2_{v+1}/v + 1]^{1/2}$-density (see (8.2.3)). Since m and s are independent for normal populations, the same is true for M and S also.

Suppose now that $u \sim N(0, 1)$ independently of (M, S). Then we get

$$\Phi\left(M \sqrt{h} \pm kS \sqrt{\frac{v+1}{v}}\right) = P\left(\frac{u - M\sqrt{h}}{S \sqrt{1 + h}} \leqq \pm k \sqrt{\frac{v+1}{v(1+h)}}\right)$$
$$= P\left(t_{v+1} \leqq \pm k \sqrt{\frac{v+1}{v(1+h)}}\right) .$$

Therefore (8.5.8) is equivalent to

$$P\left(|t_{v+1}| > k\sqrt{\frac{v+1}{v(1+h)}}\right) = \frac{1}{1+\lambda}\,.$$

From this we derive the optimal $k = k^*$ as

$$k^* = \sqrt{\frac{v(1+h)}{v+1}}\, t_{v+1,\,[2(1+\lambda)]^{-1}}\,. \tag{8.5.9}$$

Hence the V-optimal prediction interval for z^* according to (8.5.5) is

$$\delta^*(z) = (m - k^*s,\, m + k^*s) \tag{8.5.10}$$

8.5.3 Onesided infinite intervals. We may require onesided intervals such as $\delta_1(z) = (r, \infty)$ or $\delta_2(z) = (-\infty, r)$. In practice we choose linear utility functions of the following types, these being similar to (8.5.5):

$$V_1(r, z^*) = \begin{cases} \lambda(z^* - r) & [z^* < r] \\ (r - z^*) & [z^* \geqq r] \end{cases}$$

$$V_2(r, z^*) = \begin{cases} z^* - r & [z^* \leqq r] \\ \lambda(r - z^*) & [z^* > r]\,. \end{cases}$$

If $r = m - k_1 s$ or $r = m + k_2 s$ then these functions are (cf. (8.5.7))

$$V_1(k_1, z^*) = \begin{cases} \lambda(z^* - m + k_1 s) & [z^* < m - k_1 s] \\ (m - k_1 s - z^*) & [z^* \geqq m - k_1 s]\,, \end{cases} \tag{8.5.11}$$

and

$$V_2(k_2, z^*) = \begin{cases} z^* - m - k_2 s & [z^* \leqq m + k_2 s] \\ \lambda(m + k_2 s - z^*) & [z^* > m + k_2 s]\,. \end{cases} \tag{8.5.12}$$

Optimization according to Definition 8.7 then yields the V-optimal prediction intervals (being V-optimal according to (8.5.11) and (8.5.12) respectively-see TOUTENBURG (1971):

$$\delta_1^*(z) = (m - k_1^* s, \infty)$$

and

$$\delta_2^*(z) = (-\infty, m + k_2^* s) \tag{8.5.13}$$

with

$$k_1^* = k_2^* = \sqrt{\frac{v(1+h)}{v+1}}\, t_{v+1,\,(1+\lambda)^{-1}}\,. \tag{8.5.14}$$

using (8.3.2) we may apply these results to the model (8.3.1) of normal regression. We have to choose $m = \boldsymbol{x}_*'\boldsymbol{b}$ and s from (7.4.2) and put

$$\frac{v(1+h)}{v+1} = \frac{(n-p)\,(w_*^2 + \boldsymbol{x}_*'\boldsymbol{S}^{-1}\boldsymbol{x}_*)}{n-p+1}$$

Thus k^* (8.5.9) and k_1^*, k_2^* (8.5.14) are known).

8.5.4 Utility and length of intervals. We explain the problem in a special case (see Bunke (1963)). Let σ^2 known and $\theta = \mu$ be known. Then choose the statistic $z = m \sim N(\mu, \sigma^2 n^{-1})$ and the prediction interval for z^* as

$$\delta(z) = (m - k, m + k) .$$

The optimization is carried out according to the following linear utility function which depends on the length of interval explicitly:

$$V(k, z^*) = \begin{cases} z^* - m + k & [z^* \leqq m - k] \\ -2k & [m - k < z^* < m + k] \\ m + k - z^* & [z^* \geqq m + k] . \end{cases} \tag{8.5.15}$$

This leads to the expected utility

$$\begin{aligned} \overline{V}(k, \mu) = & \int_{-\infty}^{m-k} (z^* - m + k)\, p_{T^*}(z^* \mid \mu)\, \mathrm{d}z^* \\ & - 2k \int_{m-k}^{m+k} p_{T^*}(z^* \mid \mu)\, \mathrm{d}z^* + \int_{m+k}^{\infty} (m + k - z^*)\, p_{T^*}(z^* \mid \mu)\, \mathrm{d}z^* \end{aligned}$$

with the derivative

$$\begin{aligned} \frac{\partial \overline{V}(k, \mu)}{\partial k} = & \; 1 - 3P_{T^*}(m - k, m + k \mid \mu) \\ & - 2k[p_{T^*}(m + k \mid \mu) + p_{T^*}(m - k \mid \mu)] . \end{aligned} \tag{8.5.16}$$

If we define

$$\tilde{k} = k \sqrt{\frac{n}{n+1} \sigma^{-2}}$$

we get

$$P_{T^*}(m - k, m + k \mid \mu) = P\left(\left|\frac{z^* - m}{\sigma(1 + T^{-1})^{1/2}}\right| \leqq \tilde{k}\right) = 2\Phi(\tilde{k}) - 1 .$$

The equation to determine the optimal k such that

$$\int_{-\infty}^{\infty} \frac{\partial}{\partial k} \overline{V}(k, \mu)\, p_T(m \mid \mu)\, \mathrm{d}m = 0$$

takes the form (with the help of (8.5.16))

$$0 = 1 - 1.5\Phi(\tilde{k}) - \tilde{k}\varphi(\tilde{k})$$

(where φ is density of the $N(0, 1)$-distribution). Its solution is roughly

$$\tilde{k} = 0.255 .$$

This way the prediction interval for z^* which is V-optimal according to (8.5.15) is of the form

$$\delta(m) = (m - k^*, m + k^*)$$

with

$$k^* = \tilde{k}\sigma \sqrt{\frac{n+1}{n}} \approx 0.255\ \sigma \sqrt{\frac{n+1}{n}}\,.$$

8.5.5 Utility and coverage. At first we determine the expected coverage q of the interval (8.5.10) by comparing the values k^* of (8.5.9) and (8.2.9). With $\tilde{\alpha} < \alpha$

$$\sqrt{\frac{v}{v+1}}\, t_{v+1,\,\tilde{\alpha}} = t_{v,\,\alpha}$$

holds, i.e. especially

$$\sqrt{\frac{v}{v+1}}\, t_{v+1,\,[2(1+\lambda)]^{-1}} = t_{v,\,(1-g-\varepsilon^2)/2}\,. \tag{8.5.17}$$

This way the V-optimal prediction interval (8.5.10) is a $\left(\frac{\lambda}{1+\lambda} - \varepsilon^2\right)$-(coverage)-interval where ε^2 has to be determined according to (8.5.17) from tables of t_n-quantiles.

If a V-optimal prediction interval with given coverage q is required then the cost factor λ in (8.5.17) has to be calculated and the utility function V (8.5.7) must be fixed. So λ can be interpreted as a control parameter for the construction of prediction intervals with given coverage q and maximal utility (i.e. in some cases minimal length, cp. (8.5.15)).

8.6 Maximal utility and optimal tests

Let $\mathfrak{B}(Z_1, \ldots, Z_T)$ be a statistical prediction region for $\boldsymbol{Z}^*$. We define the indicator function of $\mathfrak{B}$ as

$$\Phi(\mathfrak{B}, \boldsymbol{Z}^*) = \begin{cases} 1 & \textit{for } \boldsymbol{Z}^* \in \mathfrak{B} \\ 0 & \textit{otherwise}\,. \end{cases} \tag{8.6.1}$$

With the help of this the coverage of $\mathfrak{B}$ results as

$$P_{Z^*}[\mathfrak{B}(Z_1, \ldots, Z_T)] = E_{Z^*}\Phi(\mathfrak{B}, \boldsymbol{Z}^*)$$

With that (see also Definition 8.5) $\mathfrak{B}(Z_1, \ldots, Z_T)$ is a q-prediction region for $\boldsymbol{Z}^*$ if

$$E_{ZZ^*}\Phi(\mathfrak{B}, \boldsymbol{Z}^*) = q \qquad (\text{for all } \theta \in \boldsymbol{\Omega})\,. \tag{8.6.2}$$

If a utility function $V(\mathfrak{B}, \boldsymbol{Z}^*)$ is given then condition (8.5.2) for the V-optimality of $\mathfrak{B}$ takes the form

$$\max_{\{\mathfrak{B}\}} \int\!\!\int_{ZZ^*} V(\mathfrak{B}, \boldsymbol{Z}^*)\, \mathrm{d}P_{TT^*}(\boldsymbol{Z}, \boldsymbol{Z}^* \mid \theta) = \max_{\{\mathfrak{B}\}} E_{ZZ^*}[\Phi(\mathfrak{B}, \boldsymbol{Z}^*)\, V(\mathfrak{B}, \boldsymbol{Z}^*)]\,, \tag{8.6.3}$$

where $\{\mathfrak{B}\}$ is a class of admissible prediction regions. If the optimization (8.6.3) of $\mathfrak{B}$ is carried out according to the restriction (8.6.2) of q coverage and if $V(\mathfrak{B}, \boldsymbol{Z}^*)$ is chosen to be a probability measure Q^θ over $(\mathfrak{Z}, \mathfrak{A})$ then the following theorem holds globally:

Theorem 8.6 The construction of a V-optimal q-prediction region is equivalent to the construction of an optimal (either uniformly most powerful or minimax)-test statistic $\Phi(\boldsymbol{Z}, \boldsymbol{Z}^*)$ for the test problem

$$\left.\begin{array}{l} H_0\colon (P_Z^\theta, P_{Z^*}^\theta) \\ H_1\colon (P_Z^\theta, Q^\theta) \end{array} \quad \theta \in \boldsymbol{\Omega} \right\} \tag{8.6.4}$$

given the probability of the error of first kind $(1 - q)$ (Guttman (1970), p. 36).

We shall now construct V-optimal q-prediction regions for normal populations, using assumptions (8.2.2), with normally distributed utility functions. If the statistician is interested in a prediction interval $\delta(z)$ being symmetric around μ he will choose a utility function of the form

$$Q^\theta = Q(\mu, \sigma^2) = N(\mu, \alpha^2\sigma^2) \quad [0 < \alpha < 1]\,. \tag{8.6.5}$$

This function values intervals with mean μ higher the smaller the interval length is. This way we have $\boldsymbol{Z}^* \sim N(\mu, \alpha^2\sigma^2)$ with $\alpha = 1$ under H_0 and $\alpha < 1$ under H_1. This leads to the following test problem:

$$H_0\colon \alpha = 1$$

$$H_1\colon \alpha = \alpha_1 < 1\,.$$

Based on the Neyman criterion (Witting and Nölle, 1970) we get the sufficient statistic

$$(m, s^2, z^*) \qquad \text{with} \qquad m, s^2 \text{ defined in (8.2.3)}\,.$$

According to (8.2.2) and (8.2.3) we have

$$z^* - m \sim N\big(0, (\alpha + h)\,\sigma^2\big)$$

giving the statistic

$$t = \frac{z^* - m}{s} \sim \sqrt{\alpha + h}\; t_v \qquad \text{[Theorem A32c]}\,.$$

This way we have under H_0 that

$$t \sim \sqrt{1 + h}\; t_v$$

and under H_1 that

$$t \sim \sqrt{\alpha_1 + h}\; t_v\,.$$

If $p\big(t; (\alpha + h)^{1/2}\big)$ denotes the t-density, then the uniformly most powerful test statistic $\Phi(\boldsymbol{Z}, \boldsymbol{Z}^*) = \tilde{\Phi}(t)$ is

$$\tilde{\Phi}(t) = 1 \quad \text{if} \quad \frac{p\big(t; (\alpha_1 + h)^{1/2}\big)}{p\big(t; (1 + h)^{1/2}\big)} \geqq a(q)\,, \quad \tilde{\Phi}(t) = 0\,, \quad \text{otherwise}\,, \tag{8.6.6}$$

where the constant $a(q)$ is chosen so that the probability of error of the first kind is $(1 - q)$. Along with $p(t)$ monotonically decreasing in $|t|$ the same is true for the quotient in (8.6.6) as $\alpha_1 < 1$. Therefore (8.6.6) is equivalent to

$$\tilde{\Phi}(t) = 1 \quad \text{if} \quad |t| \leqq \tilde{a}(q)\,, \quad \tilde{\Phi}(t) = 0 \quad \text{otherwise.} \tag{8.6.7}$$

The probability of the error of first kind is

$$P(|t| > \tilde{a}(q) \mid \alpha = 1) = 2P\big(\sqrt{1+h}\, t_v > \tilde{a}(q)\big) = 1 - q\,,$$
$$\big(\tilde{a}(q) = k^* \text{ from (8.2.9)}\big)\,.$$

So the q-prediction interval of z^* being uniformly most powerful according to the measure Q^θ from (8.6.5) is

$$\delta(z) = \{z^*: |t| < \tilde{a}(q)\} = (m - s\sqrt{1+h}\, t_{v,(1-q)/2}, m + s\sqrt{1+h}\, t_{v,(1-q)/2})\,.$$

The same interval has already been obtained in 8.2. Thus, the interval length is not influenced by the requirement of V-optimality according to $V = Q^\theta = N(\mu, \alpha_1^2\sigma^2)$, $\alpha_1 < 1$. (See GUTTMAN (1970) for tables of $\tilde{a}(q)$ and tests using multivariate normal distributed utility functions.)

8.7 Summary

This chapter introduces the notion of prediction intervals, and p- and (p, q)-intervals as formulated by GUTTMANN (1970). Applications to the models considered in previous chapters are given.

The notion of a utility or valuation function is introduced in Section 8.5. This enables one to express a prior preference for the prediction to lie in particular regions, and opens the door to various BAYESIAN developments. Linear utility functions are given particular prominence.

The final section considers the construction of optimal prediction regions.

9

Prediction in econometric models

9.1 Optimal prediction using random regressors

In contrast to model (4.1.1) we now assume that

$$\boldsymbol{X} \text{ is stochastic but independent of } \boldsymbol{u} \,. \tag{9.1.1}$$

The regressors $\boldsymbol{X}_i$ $(i = 1, \dots, p)$ may be regarded as generated by univariate stochastic processes $\{x_{ti}\}$ which form a multivariate (p-dimensional) stochastic process $\{x_{t1}, \dots, x_{tp}\} = \{\boldsymbol{x}(t)\}$. We assume that

$$\left.\begin{aligned} E\boldsymbol{x}(t) = \boldsymbol{g}\,, \quad E\big(\boldsymbol{x}(t) - \boldsymbol{g}\big)\big(\boldsymbol{x}(t) - \boldsymbol{g}\big)' = \boldsymbol{V} \\ \boldsymbol{V} \text{ positive definite.} \end{aligned}\right\} \tag{9.1.2}$$

If $\boldsymbol{V}$ is positive definite the process is called *regular*. According to a theorem of CRAMÉR (1966) the process is regular iff the realizations of the process do not lie with probability one in a proper subspace of E^P, that is iff every matrix $\boldsymbol{X} = (\boldsymbol{x}_1', \dots, \boldsymbol{x}_n')$ is of rank p. Then the necessary inverses exist, and in particular, the AITKEN (GLS) estimator $\boldsymbol{b} = (\boldsymbol{X}'\boldsymbol{W}^{-1}\boldsymbol{X})^{-1}\,\boldsymbol{X}'\boldsymbol{W}^{-1}\boldsymbol{y}$ can be constructed. If the same stochastic process is maintained at the prediction period T^* also, then we get from (9.1.2), putting $\boldsymbol{x}_{T*} = \boldsymbol{x}_*$, that

$$E(\boldsymbol{x}_* \mid \boldsymbol{X}) = E\boldsymbol{x}_* = \boldsymbol{g}\,,$$

and

$$E[(\boldsymbol{x}_* - \boldsymbol{g})(\boldsymbol{x}_* - \boldsymbol{g})' \mid \boldsymbol{X}] = E(\boldsymbol{x}_*\boldsymbol{x}_*' \mid \boldsymbol{X}) - \boldsymbol{g}\boldsymbol{g}' = \boldsymbol{V} = \boldsymbol{G} - \boldsymbol{g}\boldsymbol{g}'\,.$$

From this the positive definiteness of $\boldsymbol{G}$ follows. This leads to the model of generalized independent stochastic regression as

$$\left.\begin{aligned} \boldsymbol{y} = \boldsymbol{X\beta} + \boldsymbol{u}\,, \quad \textit{rank}\; \boldsymbol{X} = p \\ E(\boldsymbol{u} \mid \boldsymbol{X}) = E\boldsymbol{u} = \boldsymbol{0}\,, \quad E(\boldsymbol{uu}' \mid \boldsymbol{X}) = E\boldsymbol{uu}' = \sigma^2\boldsymbol{W}\,. \end{aligned}\right\} \tag{9.1.3}$$

We assume analogously to (5.1.2) and (5.1.3) that the prediction index T^* satisfies

$$\boldsymbol{y}_* = \boldsymbol{X}_*\boldsymbol{\beta} + \boldsymbol{u}_*\,, \quad E\boldsymbol{u}_* = \boldsymbol{0}\,, \quad E\boldsymbol{u}_*\boldsymbol{u}_*' = \sigma^2\boldsymbol{W}_*\,, \quad E\boldsymbol{u}\boldsymbol{u}_*' = \sigma^2\boldsymbol{W}_0\,. \tag{9.1.4}$$

We set up the linear predictor (see (5.2.1)) as

$$\boldsymbol{p} = \boldsymbol{C}'\boldsymbol{y} + \boldsymbol{d}\,.$$

In particular we may choose in connection with the given stochastic model

$$C' = X_* \tilde{C}' .$$

We may use results from the model (4.1.1) with nonstochastic regressors and generalize them to the new model [(9.1.3), (9.1.4)] if we choose the conditional risk function

$$\tilde{R}(\boldsymbol{p}) = E[(\boldsymbol{p} - \boldsymbol{y}_*)' \boldsymbol{A}(\boldsymbol{p} - \boldsymbol{y}_*) \mid \boldsymbol{X}, \boldsymbol{X}_*] \tag{9.1.5}$$

and unbiasedness in the sense that

$$E[(\boldsymbol{p} - \boldsymbol{y}_*) \mid \boldsymbol{X}, \boldsymbol{X}_*] = 0 \quad \text{(for all } \boldsymbol{\beta}, \sigma^2; \boldsymbol{X}, \boldsymbol{X}_* \text{ fixed)} . \tag{9.1.6}$$

At first we give attention to the following fact. Let $\boldsymbol{p}_1$ and $\boldsymbol{p}_2$ be two predictors with nonstochastic regressors and let $\boldsymbol{p}_1$ be *R-better* than $\boldsymbol{p}_2$. If the regressors are now stochastic this relation may also be written as

$$\tilde{R}(\boldsymbol{p}_2) - \tilde{R}(\boldsymbol{p}_1) \geqq 0 .$$

Integrating over $\boldsymbol{X}$ and $\boldsymbol{X}_*$, we see that the unconditional risk $R = E(\boldsymbol{p} - \boldsymbol{y}_*)' \boldsymbol{A}(\boldsymbol{p} - \boldsymbol{y}_*)$ satisfies

$$R(\boldsymbol{p}_2) - R(\boldsymbol{p}_1) \geqq 0 .$$

In other words the relative goodness of the two predictors in the nonstochastic model (4.1.1) stay valid in the stochastic model (9.1.3). In particular the $\tilde{R}$-optimal predictions in the model (9.1.4) and the R-optimal predictions in the model (4.1.1) coincide.

Otherwise, if the general risk function

$$R(\boldsymbol{p}) = E(\boldsymbol{p} - \boldsymbol{y}_*)' \boldsymbol{A}(\boldsymbol{p} - \boldsymbol{y}_*)$$

is given which requires expectation over the regressors as well as the error process, the calculation of R-optimal predictors becomes difficult. These difficulties cannot be overcome without further strong assumptions upon the process $\{x_t\}$. In contrast to this the following risk function

$$R_x(\boldsymbol{p}) = E[(\boldsymbol{p} - \boldsymbol{y}_*)' \boldsymbol{A}(\boldsymbol{p} - \boldsymbol{y}_*) \mid \boldsymbol{X}] \tag{9.1.7}$$

leads to tractable mathematics. In the case $n = 1$ we get unbiased heterogeneous R_x-optimal predictions in the class $\{\boldsymbol{p} = \boldsymbol{x}'_* \boldsymbol{C}' \boldsymbol{y}\}$ according to the following theorem, which is due to TOUTENBURG (1970a).

Theorem 9.1 In the class $\{\boldsymbol{p} : \boldsymbol{p} = \boldsymbol{x}'_* \tilde{\boldsymbol{C}}' \boldsymbol{y}\}$, the R_x-optimal predictions of a component $y_{\tau *}$ of $\boldsymbol{y}_*$ (9.1.4) are the following.

(a) Biased predictors

$$\bar{p}_1 = \boldsymbol{x}'_* \hat{\boldsymbol{\beta}}_2 + \boldsymbol{x}'_* \boldsymbol{G}^{-1} \boldsymbol{g} \boldsymbol{w}' \boldsymbol{B}^{-1} \boldsymbol{y}$$

where

$$\boldsymbol{B} = \sigma^{-2} \boldsymbol{X} \boldsymbol{\beta} \boldsymbol{\beta}' \boldsymbol{X}' + \boldsymbol{W} , \qquad \hat{\boldsymbol{\beta}}_2 = \sigma^{-2} \boldsymbol{\beta} \boldsymbol{\beta}' \boldsymbol{X}' \boldsymbol{B}^{-1} \boldsymbol{y} .$$

(b) Unbiased predictors (in the sense $E[(\boldsymbol{p} - \boldsymbol{y}_*) \mid \boldsymbol{X}] = \boldsymbol{0}$)

$$\bar{p}_2 = \bar{p}_1 - \boldsymbol{x}'_1 \boldsymbol{G}^{-1} \boldsymbol{g} \boldsymbol{w}' \boldsymbol{B}^{-1} \boldsymbol{X} (\boldsymbol{X}' \boldsymbol{B}^{-1} \boldsymbol{X})^{-1} \boldsymbol{X}' \boldsymbol{B}^{-1} \boldsymbol{y} .$$

As both predictors contain the unknown vector $\sigma^{-1}\boldsymbol{\beta}$ the use of prior information is necessary to make them practicable. These problems were discussed in detail in 5.5. and in TOUTENBURG (1970a) and hence will be omitted here.

Instead we shall erect a new prediction set-up in order to overcome these difficulties and to get practicable predictors. As $\boldsymbol{x}_*$ is the (unknown) realization of the process $\{x_t\}$ at the index T^* the following set up should be taken:

$$p = \hat{\boldsymbol{x}}'_* \boldsymbol{C}' \boldsymbol{y}$$

where $\hat{\boldsymbol{x}}'_* = f(\boldsymbol{X})$ shall be an estimator of $\boldsymbol{x}^*$ on the basis of the realizations $\{x_1\}, \ldots, \{x_n\}$. Note especially that $\hat{\boldsymbol{x}}'_*$ should not depend on $\boldsymbol{y}$ so that

$$E(\hat{\boldsymbol{x}}'_* \mid \boldsymbol{X}) = \hat{\boldsymbol{x}}'_* .$$

On account of

$$p - y_* = (\hat{\boldsymbol{x}}'_* \boldsymbol{C}' \boldsymbol{X} - \boldsymbol{x}'_*)\, \boldsymbol{\beta} + \hat{\boldsymbol{x}}'_* \boldsymbol{C}' \boldsymbol{u} - u_*$$

and

$$E(p - y_* \mid \boldsymbol{X}) = (\hat{\boldsymbol{x}}'_* \boldsymbol{C}' \boldsymbol{X} - \boldsymbol{g}')\, \boldsymbol{\beta} ,$$

the conditional unbiasedness of p requires the following equivalent condition

$$\hat{\boldsymbol{x}}'_* \boldsymbol{C}' \boldsymbol{X} - \boldsymbol{g}' = \boldsymbol{0}' . \tag{9.1.10}$$

From this we get the optimal $\boldsymbol{C}$ as the solution of the optimization problem

$$\min_{\boldsymbol{C}} \{R_x(p) - 2(\hat{\boldsymbol{x}}'_* \boldsymbol{C}' \boldsymbol{X} - \boldsymbol{g}')\, \lambda\} ,$$

where according to (9.1.2) and (9.1.7) the following result holds

$$R_x(p) = E[(p - y_*)^2 \mid \boldsymbol{X}] = \sigma^2 \hat{\boldsymbol{x}}'_* \boldsymbol{C}' \boldsymbol{W} \boldsymbol{C} \hat{\boldsymbol{x}}_* - 2\sigma^2 \hat{\boldsymbol{x}}'_* \boldsymbol{C}' \boldsymbol{w} + \sigma_*^2 + \boldsymbol{\beta}' \boldsymbol{V} \boldsymbol{\beta} .$$

Theorem 9.2 In the class of admissible predictors

$$\{p : p = \hat{\boldsymbol{x}}'_* \boldsymbol{C}' \boldsymbol{y} \mid E(\hat{\boldsymbol{x}}'_* \mid \boldsymbol{X}) = \hat{\boldsymbol{x}}'_*\}$$

the *R_x-optimal conditional unbiased predictor* of a component $y_{\tau*}$ of $\boldsymbol{y}_*$ (9.1.4) is

$$\bar{p}_3 = \boldsymbol{g}' \boldsymbol{b} + \boldsymbol{w}' \boldsymbol{W}^{-1} (\boldsymbol{y} - \boldsymbol{X} \boldsymbol{b}) \tag{9.1.11}$$

with

$$E(\bar{p}_3 - y_* \mid \boldsymbol{X}) = \boldsymbol{0}$$

and

$$R_x(\bar{p}_3) = \boldsymbol{\beta}' \boldsymbol{V} \boldsymbol{\beta} + \sigma^2 (\boldsymbol{g}' - \boldsymbol{w}' \boldsymbol{W}^{-1} \boldsymbol{X})\, \boldsymbol{S}^{-1} (\boldsymbol{g}' - \boldsymbol{w}' \boldsymbol{W}^{-1} \boldsymbol{X})' + R(\hat{p}_1) . \tag{9.1.12}$$

(In this expression $R(\hat{p}_1)$ is taken from (5.2.10).)

In practice $\boldsymbol{g}$ is estimated by extrapolation of the process $\{x_t\}$ using e. g. autoregressive techniques, or by the sample mean.

9.2 Prediction in multivariate regression

In preparing the prediction of endogeneous variables in econometric models we at first regard the multivariate regular regression model (here $\tilde{p} = \sum_{m=1}^{M} p_m$)

$$\left.\begin{array}{c} \boldsymbol{y} = \boldsymbol{Z}\beta + \boldsymbol{u}\,, \quad E\boldsymbol{u} = \boldsymbol{0}\,, \quad E\boldsymbol{u}\boldsymbol{u}' = \sigma^2\boldsymbol{\Phi} \\ \boldsymbol{Z} \text{ a nonstochastic matrix of order } Mn \times \tilde{p},\ rank\ \boldsymbol{Z} = \tilde{p} \end{array}\right\} \tag{9.2.1}$$

where $\boldsymbol{Z} = diag\,(\boldsymbol{X}_1, \ldots, \boldsymbol{X}_M)$ is the *'system matrix'* of regressors derived from $diag\,(\boldsymbol{X}, \ldots, \boldsymbol{X})$ by cancelling some $\boldsymbol{X}_i'$s according to the given null restrictions. If we remember Theorem 6.4 (cp. $\hat{\boldsymbol{p}}_6$ and $\boldsymbol{b}_6$ for $\boldsymbol{V} = \boldsymbol{0}$) the null restrictions ensure a gain in efficiency in estimating and predicting. We assume that

$$\begin{pmatrix} \boldsymbol{y}_{1*} \\ \vdots \\ \boldsymbol{y}_{M*} \end{pmatrix} = \begin{pmatrix} \boldsymbol{x}_{1*}' & \cdots & \boldsymbol{0} \\ \vdots & & \vdots \\ \boldsymbol{0} & \cdots & \boldsymbol{x}_{M*}' \end{pmatrix} \begin{pmatrix} \beta_1 \\ \vdots \\ \beta_M \end{pmatrix} + \begin{pmatrix} \boldsymbol{u}_{1*} \\ \vdots \\ \boldsymbol{u}_{M*} \end{pmatrix}$$

holds at index T^* analogously to the sample model. In matrix notation this is

$$\left.\begin{array}{l} \boldsymbol{y}_* = \boldsymbol{Z}_*\beta + \boldsymbol{u}_*\,, \quad E\boldsymbol{u}_* = \boldsymbol{0}\,, \quad E\boldsymbol{u}_*\boldsymbol{u}_*' = \sigma^2\boldsymbol{W}_* \\ E\boldsymbol{u}\boldsymbol{u}_*' = \sigma^2\boldsymbol{W}_0 = \sigma^2(\boldsymbol{w}_1, \ldots, \boldsymbol{w}_M), \boldsymbol{Z}_* \text{ nonstochastic}\,. \end{array}\right\} \tag{9.2.2}$$

The prediction model follows the same null restrictions as the sample model (this is necessary to enable us to predict $\boldsymbol{y}_*$ on the basis of the model (9.2.1)).

We confine ourselves to the homogeneous set-up with $\boldsymbol{C}$ a $Mn \times M$-matrix

$$\boldsymbol{p} = \boldsymbol{C}'\boldsymbol{y} = \begin{pmatrix} \boldsymbol{c}_1' \\ \vdots \\ \boldsymbol{c}_M' \end{pmatrix} \boldsymbol{y} = \begin{pmatrix} \boldsymbol{c}_1'\boldsymbol{y} \\ \vdots \\ \boldsymbol{c}_M'\boldsymbol{y} \end{pmatrix} = \begin{pmatrix} p_1 \\ \vdots \\ p_M \end{pmatrix}. \tag{9.2.3}$$

The component $p_m = \boldsymbol{c}_m'\boldsymbol{y}$ takes into account all the sample information contained in $\boldsymbol{y}$. We require the minimization of the risk

$$R(\boldsymbol{p}) = E(\boldsymbol{p} - \boldsymbol{y}_*)'\,\boldsymbol{I}(\boldsymbol{p} - \boldsymbol{y})_*$$

where the choice $\boldsymbol{A} = \boldsymbol{I}$ expresses the symmetric treatment of all M regressands in (9.2.1). This leads to

$$R(\boldsymbol{p}) = \sum_{m=1}^{M} R(p_m) = \sum_{m=1}^{M} E(p_m - y_{m*})^2\,.$$

If we require the unbiasedness of $\boldsymbol{p}$ in addition, i.e.

$$\boldsymbol{C}'\boldsymbol{Z} - \boldsymbol{Z}_* = 0\,,$$

then the R-optimal prediction is derived from Theorem 5.2 as

$$\hat{\boldsymbol{p}}_3 = \boldsymbol{Z}_* \boldsymbol{b} + \boldsymbol{W}_0' \boldsymbol{\Phi}^{-1}(\boldsymbol{y} - \boldsymbol{Z}\boldsymbol{b}) \tag{9.2.4}$$

where $\boldsymbol{b}$ is the GLSE. Its risk will be

$$R(\hat{\boldsymbol{p}}_3) = R(\hat{\boldsymbol{p}}_1) + tr\,[(\boldsymbol{Z}_* - \boldsymbol{W}_0'\boldsymbol{W}^{-1}\boldsymbol{Z})\,\boldsymbol{V}_b(\boldsymbol{Z}_* - \boldsymbol{W}_0'\boldsymbol{W}^{-1}\boldsymbol{Z})'] \tag{9.2.5}$$

with

$$R(\hat{\boldsymbol{p}}_1) = \sigma^2\, tr\,(\boldsymbol{W}_* - \boldsymbol{W}_0'\boldsymbol{\Phi}^{-1}\boldsymbol{W}_0)\,,\quad \boldsymbol{b} = (\boldsymbol{Z}'\boldsymbol{\Phi}^{-1}\boldsymbol{Z})^{-1}\,\boldsymbol{Z}'\boldsymbol{\Phi}^{-1}\boldsymbol{y}\,,$$
$$\boldsymbol{V}_b = \sigma^2(\boldsymbol{Z}'\boldsymbol{\Phi}^{-1}\boldsymbol{Z})^{-1}\,.$$

We now turn to investigate the influence of the system information (contained in $\boldsymbol{y}$ or in $\sigma^2\boldsymbol{\Phi}$) on the goodness of prediction, as developed by Toutenburg (1970c). Without loss of generality let us assume a two-equation system ($M = 2$) of classical type:

$$\left.\begin{aligned} \boldsymbol{y}_1 &= \boldsymbol{X}_1\boldsymbol{\beta}_1 + \boldsymbol{u}_1 \\ \boldsymbol{y}_2 &= \boldsymbol{X}_2\boldsymbol{\beta}_2 + \boldsymbol{u}_2 \end{aligned}\right\}\ \boldsymbol{y} = \boldsymbol{Z}\boldsymbol{\beta} + \boldsymbol{u}\,,\quad E\,\boldsymbol{u} = \boldsymbol{0}\,,$$

$$E\boldsymbol{u}\boldsymbol{u}' = \sigma^2\boldsymbol{\Phi} = \sigma^2\begin{pmatrix} \sigma_1^2 & \varrho\sigma_1\sigma_2 \\ \varrho\sigma_1\sigma_2 & \sigma_2^2 \end{pmatrix} \otimes \boldsymbol{I} \tag{9.2.6}$$

($\otimes$ denotes the Kronecker product, cp. A 39). Here the errors $\boldsymbol{u}_1$ and $\boldsymbol{u}_2$ are homoscedastic with covariance matrix $\sigma^2\boldsymbol{W}_{11} = \sigma^2\sigma_1^2\boldsymbol{I}$ and $\sigma^2\boldsymbol{W}_{22} = \sigma^2\sigma_2^2\boldsymbol{I}$, respectively. Since the error vectors are correlated, the dependent variables are also

$$E\boldsymbol{u}_1\boldsymbol{u}_2' = \sigma^2\boldsymbol{W}_{12} = \sigma^2\varrho\sigma_1\sigma_2\boldsymbol{I}\,.$$

The prediction model is

$$\left.\begin{aligned} y_{1*} &= \boldsymbol{x}_{1*}'\boldsymbol{\beta}_2 + u_{1*} \\ y_{2*} &= \boldsymbol{x}_{2*}'\boldsymbol{\beta}_2 + u_{2*} \end{aligned}\right\}\ \boldsymbol{y}_* = \boldsymbol{Z}_*\boldsymbol{\beta} + \boldsymbol{u}_*\,,\qquad E\boldsymbol{u}_* = \boldsymbol{0}\,,$$

$$E\,\boldsymbol{u}_*\boldsymbol{u}_*' = \sigma^2\boldsymbol{W}_* = \sigma^2\begin{pmatrix} \sigma_1^2 & \varrho\sigma_1\sigma_2 \\ \varrho\sigma_1\sigma_2 & \sigma_2^2 \end{pmatrix} \tag{9.2.7}$$

(i.e. $\boldsymbol{\Phi} = \boldsymbol{W}_* \otimes \boldsymbol{I}$ if we keep in mind (9.2.6) and (9.2.7)).

As the errors u_m are homoscedastic or temporally uncorrelated we have $E[u_m u_{m*}] = 0$ and $E[u_m u_{m*}] = 0$ ($m \neq m'$), and therefore $\boldsymbol{W}_0 = \boldsymbol{0}$. By minimizing the risk $R(p_m) = E(p_m - y_{m*})^2$, we get the homogeneous R-optimal partial predictions of each of the y_{m*} according to Theorem 5.3 as $\check{p}_m$ where

$$\check{p}_m = \boldsymbol{x}_{m*}'\check{\boldsymbol{b}}_m \qquad (m = 1, 2) \tag{9.2.8}$$

with

$$\check{\boldsymbol{b}}_m = (\boldsymbol{X}_m'\boldsymbol{X}_m)^{-1}\,\boldsymbol{X}_m'\boldsymbol{y}_m\,,$$

and

$$R(\check{p}_m) = \sigma^2\sigma_m^2 + \sigma^2\sigma_m^2\boldsymbol{x}_{m*}'(\boldsymbol{X}_m'\boldsymbol{X}_m)^{-1}\,\boldsymbol{x}_{m*} \tag{9.2.9}$$

The partial prediction of the whole vector $\boldsymbol{y}'_* = (y_{1*}, y_{2*})$ is therefore

$$\check{\boldsymbol{p}}_3 = \boldsymbol{Z}_* \begin{pmatrix} \check{\boldsymbol{b}}_1 \\ \check{\boldsymbol{b}}_2 \end{pmatrix} = \boldsymbol{Z}_* \check{\boldsymbol{b}} \,. \tag{9.2.10}$$

We have

$$R(\check{\boldsymbol{p}}_3) = \sigma^2 \, tr \, \boldsymbol{W}_* + tr \, \boldsymbol{Z}_* \begin{pmatrix} \sigma^2 \sigma_1^2 (\boldsymbol{X}_1' \boldsymbol{X}_1)^{-1} & \boldsymbol{0} \\ \boldsymbol{0} & \sigma^2 \sigma_1^2 (\boldsymbol{X}_2' \boldsymbol{X}_2)^{-1} \end{pmatrix} \boldsymbol{Z}'_* \tag{9.2.11}$$

In the special case $\boldsymbol{W}_0 = \boldsymbol{0}$, the system prediction $\hat{\boldsymbol{p}}_3$ (9.2.4) yields

$$\hat{\boldsymbol{p}}_3 = \boldsymbol{Z}_* \boldsymbol{b}$$

with

$$R(\hat{\boldsymbol{p}}_3) = \sigma^2 \, tr \, \boldsymbol{W}_* + tr \, \boldsymbol{Z}_* \begin{pmatrix} \boldsymbol{V}_1 & \boldsymbol{0} \\ \boldsymbol{0} & \boldsymbol{V}_2 \end{pmatrix} \boldsymbol{Z}'_* \,. \tag{9.2.12}$$

Here $\boldsymbol{V}_1$, $\boldsymbol{V}_2$ are the covariance matrices of the components $\boldsymbol{b}_1$, $\boldsymbol{b}_2$ of the system estimator $\boldsymbol{b} = (\boldsymbol{Z}'\boldsymbol{\Phi}^{-1}\boldsymbol{Z})^{-1} \boldsymbol{Z}'\boldsymbol{\Phi}^{-1}\boldsymbol{y}$. From (9.2.6) we deduce that

$$\begin{aligned} \boldsymbol{\Phi}^{-1} &= (1 - \varrho^2)^{-1} \begin{pmatrix} \sigma_1^{-2} & -\varrho \sigma_1^{-1} \sigma_2^{-1} \\ -\varrho \sigma_1^{-1} \sigma_2^{-1} & \sigma_2^{-2} \end{pmatrix} \otimes \boldsymbol{I} \\ &= \begin{pmatrix} a_1 & a_2 \\ a_2 & a_3 \end{pmatrix} \times \boldsymbol{I} \,. \end{aligned}$$

Therefore from (6.5.17)

$$\begin{aligned} \boldsymbol{V}_2 &= \sigma^2 a_3^{-1} [\boldsymbol{X}_2' \boldsymbol{X}_2 - \varrho^2 \boldsymbol{X}_2' \boldsymbol{X}_1 (\boldsymbol{X}_1' \boldsymbol{X}_1)^{-1} \boldsymbol{X}_1' \boldsymbol{X}_2]^{-1} \\ &= \sigma^2 \sigma_2^2 [\boldsymbol{X}_2' \boldsymbol{X}_2 + (1 - \varrho^2)^{-1} \varrho^2 (\boldsymbol{X}_2' \boldsymbol{X}_2 - \boldsymbol{X}_2' \boldsymbol{X}_1 (\boldsymbol{X}_1' \boldsymbol{X}_1)^{-1} \boldsymbol{X}_1' \boldsymbol{X}_2)]^{-1} \,, \end{aligned}$$

and

$$\begin{aligned} \boldsymbol{V}_1 &= \sigma^2 a_1^{-1} (\boldsymbol{X}_1' \boldsymbol{X}_1)^{-1} + a_2^2 a_1^{-2} (\boldsymbol{X}_1' \boldsymbol{X}_1)^{-1} \boldsymbol{X}_1' \boldsymbol{X}_2 \boldsymbol{V}_2 \boldsymbol{X}_2' \boldsymbol{X}_1 (\boldsymbol{X}_1' \boldsymbol{X}_1)^{-1} \\ &= \sigma^2 \sigma_1^2 (1 - \varrho^2) (\boldsymbol{X}_1' \boldsymbol{X}_1)^{-1} + \varrho^2 \sigma_2^{-2} \sigma_1^2 (\boldsymbol{X}_1' \boldsymbol{X}_1)^{-1} \boldsymbol{X}_1' \boldsymbol{X}_2 \boldsymbol{V}_2 \boldsymbol{X}_2' \boldsymbol{X}_1 (\boldsymbol{X}_1' \boldsymbol{X}_1)^{-1} \,. \end{aligned}$$

In comparing the predictors $\check{\boldsymbol{p}}_3$ and $\hat{\boldsymbol{p}}_3$ we shall need the following results:

$$\boldsymbol{M}_1 = \boldsymbol{I} - \boldsymbol{X}_1 (\boldsymbol{X}_1' \boldsymbol{X}_1)^{-1} \boldsymbol{X}_1' \quad \text{is idempotent} \,,$$

and

$$\begin{aligned} \boldsymbol{A} &= \boldsymbol{X}_2' \boldsymbol{X}_2 - \boldsymbol{X}_2' \boldsymbol{X}_1 (\boldsymbol{X}_1' \boldsymbol{X}_1)^{-1} \boldsymbol{X}_1' \boldsymbol{X}_2 \\ &= (\boldsymbol{X}_2' \boldsymbol{M}_1)(\boldsymbol{X}_2' \boldsymbol{M}_1)' \quad \text{is nonnegative definite} \,, \end{aligned}$$

which with (A11) allows us to deduce that

$$\boldsymbol{V}_2^{-1} = \sigma^{-2} \sigma_2^{-2} (\boldsymbol{X}_2' \boldsymbol{X}_2) + \sigma^{-2} \sigma_2^{-2} (1 - \varrho^2)^{-1} \varrho^2 \boldsymbol{A} \,.$$

Furthermore we may derive the following results

$$\sigma^2 \sigma_2^2 (\boldsymbol{X}_2' \boldsymbol{X}_2)^{-1} - \boldsymbol{V}_2 \quad \text{is nonnegative definite} \tag{9.2.13}$$

$$\boldsymbol{M}_2 = \boldsymbol{I} - \boldsymbol{X}_2 (\boldsymbol{X}_2' \boldsymbol{X}_2)^{-1} \boldsymbol{X}_2' \quad \text{is idempotent} \,,$$

$$\boldsymbol{B}^{-1} = \boldsymbol{X}_2' \boldsymbol{X}_2 + (1 - \varrho^2)^{-1} \varrho^2 \boldsymbol{A} \,,$$

$$(\boldsymbol{X}_2' \boldsymbol{X}_2)^{-1} - \boldsymbol{B} = \boldsymbol{C}' \boldsymbol{C} \quad \text{is nonnegative definite [cp. A11]} \,,$$

and

$$\sigma^2\sigma_2^2(\boldsymbol{X}_1'\boldsymbol{X}_1)^{-1} - \boldsymbol{V}_1 = \sigma^2\sigma_1^2\varrho^2(\boldsymbol{X}_1'\boldsymbol{X}_1)^{-1}\,\boldsymbol{X}_1'[\boldsymbol{I} - \boldsymbol{X}_2\boldsymbol{B}\boldsymbol{X}_2')]\,\boldsymbol{X}_1(\boldsymbol{X}_1'\boldsymbol{X}_1)^{-1}$$
$$- \sigma^2\sigma_1^2\varrho^2(\boldsymbol{X}_1'\boldsymbol{X}_1)^{-1}\,\boldsymbol{X}_1'[\boldsymbol{M}_2\boldsymbol{M}_2 + \boldsymbol{X}_2\boldsymbol{C}'\boldsymbol{C}\boldsymbol{X}_2']$$
$$\times \boldsymbol{X}_1(\boldsymbol{X}_1'\boldsymbol{X}_1)^{-1} \tag{9.2.14}$$

is nonnegative definite (cp. A9).

Using (9.2.11)—(9.2.14) we now have

$$R(\check{\boldsymbol{p}}_3) - R(\hat{\boldsymbol{p}}_3) = tr\,\boldsymbol{Z}_* \begin{pmatrix} \sigma^2\sigma_1^2(\boldsymbol{X}_1'\boldsymbol{X}_1)^{-1} - \boldsymbol{V}_1 & \boldsymbol{0} \\ \boldsymbol{0} & \sigma^2\sigma_2^2(\boldsymbol{X}_2'\boldsymbol{X}_2)^{-1} - \boldsymbol{V}_2 \end{pmatrix} \boldsymbol{Z}_*' \geqq 0\,. \tag{9.2.15}$$

Since $\boldsymbol{Z}_* \neq \boldsymbol{0}$ is assumed, the equality in (9.2.15) obtains iff $\varrho = 0$ or $\boldsymbol{X}_1 = \boldsymbol{X}_2 = \boldsymbol{X}$. From this it follows that $\sigma_1^{-2}\boldsymbol{V}_1 = \sigma_2^{-2}\boldsymbol{V}_2 = \sigma^2(\boldsymbol{X}'\boldsymbol{X})^{-1}$.

If $\varrho = 0$ then the individual equations are uncorrelated. In this case there is no real system but only a set of unrelated regressions which may be estimated individually. Analogously the case $\boldsymbol{X}_1 = \boldsymbol{X}_2 = \boldsymbol{X}$ means that 9.2.6) is only *one* regression model $\boldsymbol{y} = \boldsymbol{X}\boldsymbol{\beta} + \boldsymbol{u}$ with $E\boldsymbol{u}\boldsymbol{u}' = \sigma^2\boldsymbol{I}$ (i.e. $\sigma_1 = \sigma_2$). The above results may be summarized in the following theorem.

Theorem 9.3 In the multivariate regression model, the system prediction $\hat{\boldsymbol{p}}_3$ is R-better than the partial prediction $\check{\boldsymbol{p}}_3$ which is set up from the single equation predictors $\check{p}_m$ $(m = 1, 2)$ and which does not use the system information contained in the covariance matrix. If $\varrho = 0$ or $\boldsymbol{X}_1 = \boldsymbol{X}_2$ is fulfilled we have $\boldsymbol{b} = \check{\boldsymbol{b}}$ and therefore $\hat{\boldsymbol{p}}_3 = \check{\boldsymbol{p}}_3$.

9.3 Classical multivariate regression

In the multivariate model (9.2.1) the regressors were assumed to be nonstochastic, i.e. $\boldsymbol{Z}$ was fixed in repeated sampling. Now we leave this restriction and work with stochastic regressors but assume on the processes $\{x(t)\}$, $\{u(t)\}$ a weaker condition than in (9.1.3) where $E(\boldsymbol{u} \mid \boldsymbol{X}) = \boldsymbol{0}$ was required. We define the multivariate (classical) contemporaneously independent regression model by the following equations:

$$\boldsymbol{y}_m = \boldsymbol{X}\boldsymbol{\beta}_m + \boldsymbol{u}_m \qquad (m = 1, \dots, M) \tag{9.3.1}$$

$$E\boldsymbol{u}_m = \boldsymbol{0}\,, \qquad E\boldsymbol{u}_m\boldsymbol{u}_{m'}' = w_{mm'}\boldsymbol{I} \qquad (m, m' = 1, \dots, M) \tag{9.3.2}$$

$$plim\; n^{-1}\boldsymbol{u}_m'\boldsymbol{u}_m = w_{mm'} \tag{9.3.3}$$

$$E\boldsymbol{x}(t)\,\boldsymbol{x}'(t) = \boldsymbol{\Sigma}_{xx} \qquad \text{(positive definite)} \tag{9.3.4}$$

$$plim\; n^{-1}\boldsymbol{X}'\boldsymbol{u}_m = \boldsymbol{0} \qquad (m = 1, \dots, M) \tag{9.3.5}$$

and

$$\lim_{n\to\infty} E n^{-1}\boldsymbol{X}'\boldsymbol{X} = \boldsymbol{\Sigma}_{xx} . \tag{9.3.6}$$

If we define

$$\boldsymbol{y}' = (\boldsymbol{y}_1', \dots, \boldsymbol{y}_M') , \quad \boldsymbol{u}' = (\boldsymbol{u}_1', \dots, \boldsymbol{u}_M') ,$$
$$\boldsymbol{Z} = diag\,(\boldsymbol{X}, \dots, \boldsymbol{X}) , \quad \boldsymbol{\beta}' = (\boldsymbol{\beta}_1', \dots, \boldsymbol{\beta}_M') ,$$
$$\boldsymbol{W} = (w_{mm'})$$

we get the model as

$$\left.\begin{aligned} \boldsymbol{y} = \boldsymbol{Z}\boldsymbol{\beta} + \boldsymbol{u} , \quad E\boldsymbol{u} = \boldsymbol{0} \\ E\boldsymbol{u}\boldsymbol{u}' = \boldsymbol{W} \otimes \boldsymbol{I} \quad (\boldsymbol{W} \text{ positive definite}) \\ plim\, n^{-1} \begin{pmatrix} tr\, \boldsymbol{u}_1\boldsymbol{u}_1' & \cdots & tr\, \boldsymbol{u}_1\boldsymbol{u}_M' \\ \vdots & & \vdots \\ tr\, \boldsymbol{u}_M\boldsymbol{u}_1' & \cdots & tr\, \boldsymbol{u}_M\boldsymbol{u}_M' \end{pmatrix} = \boldsymbol{W} \\ plim\, n^{-1}\boldsymbol{Z}'\boldsymbol{u} = \boldsymbol{0} , \quad plim\, n^{-1}\boldsymbol{Z}'\boldsymbol{Z} = \boldsymbol{\Sigma}_{xx} \otimes \boldsymbol{I} . \end{aligned}\right\} \tag{9.3.7}$$

The assumption $plim\, T^{-1}\boldsymbol{Z}\boldsymbol{u} = \boldsymbol{0}$ is the formulation of the contemporaneously uncorrelatedness of the stochastic processes $\{x(t)\}$ and $\{u(t)\}$. This ensures the possibility of deriving consistent estimators of the parameters. As $\boldsymbol{Z}$ fulfills the condition $\boldsymbol{X}_1 = \cdots = \boldsymbol{X}_M = \boldsymbol{X}$ we can use the results of Theorem 9.3 and give the statement that $\boldsymbol{b} = \check{\boldsymbol{b}}$, i.e. the parameter vectors $\boldsymbol{\beta}_m$ may be estimated partially from the m-th equation $\boldsymbol{y}_m = \boldsymbol{X}\boldsymbol{\beta}_m + \boldsymbol{u}_m$. From [A 38] we get

$$\check{\boldsymbol{b}}_m = (\boldsymbol{X}'\boldsymbol{X})^{-1}\, \boldsymbol{X}'\boldsymbol{y}_m , \tag{9.3.8}$$

$$plim\, \check{\boldsymbol{b}}_m = \boldsymbol{\beta}_m = \overline{E}\check{\boldsymbol{b}}_m . \tag{9.3.9}$$

In other words, $\boldsymbol{b}_m$ is consistent and asymptotically unbiased and has the asymptotic covariance matrix (see A38)

$$\overline{\boldsymbol{V}}(\check{\boldsymbol{b}}_m, \check{\boldsymbol{b}}_{m'}) = n^{-1} w_{mm'}\, \boldsymbol{\Sigma}_{xx}^{-1} . \tag{9.3.10}$$

The elements $w_{mm'}$ of $\boldsymbol{W}$ are estimated as usual by the consistent estimators

$$\overline{w}_{mm'} = (n-p)^{-1}\, (\boldsymbol{y}_m - \boldsymbol{X}\check{\boldsymbol{b}}_m)'\, (\boldsymbol{y}_{m'} - \boldsymbol{X}\boldsymbol{b}_{m'})' , \tag{9.3.11}$$

Now we can estimate $\overline{\boldsymbol{V}}(\check{\boldsymbol{b}}_m, \check{\boldsymbol{b}}_{m'})$ consistently by

$$\boldsymbol{S}(\check{\boldsymbol{b}}_m, \check{\boldsymbol{b}}_{m'}) = \overline{w}_{mm'}(\boldsymbol{X}'\boldsymbol{X})^{-1} . \tag{9.3.12}$$

This leads to the following theorem.

Theorem 9.4 In the multivariate (classical) contemporaneously uncorrelated regression model (9.3.7) the classical OLS-estimates

$$\check{\boldsymbol{b}} = \begin{pmatrix} \check{\boldsymbol{b}}_1 \\ \vdots \\ \check{\boldsymbol{b}}_M \end{pmatrix} = (\boldsymbol{Z}'\boldsymbol{Z})^{-1}\, \boldsymbol{Z}'\boldsymbol{y} , \quad \overline{\boldsymbol{W}} = (n-p)^{-1}\, (\boldsymbol{y} - \boldsymbol{Z}\check{\boldsymbol{b}})'\, (\boldsymbol{y} - \boldsymbol{Z}\check{\boldsymbol{b}}) ,$$

$$\boldsymbol{S}(\check{\boldsymbol{b}}, \check{\boldsymbol{b}}) = \overline{\boldsymbol{W}} \otimes (\boldsymbol{X}'\boldsymbol{X})^{-1}$$

are consistent and asymptotically unbiased. Also

$$plim\ \check{\boldsymbol{b}} = \boldsymbol{\beta}\ , \qquad plim\ \overline{\boldsymbol{W}} = \boldsymbol{W}\ ,$$

$$\overline{\boldsymbol{V}}(\check{\boldsymbol{b}}, \check{\boldsymbol{b}}) = \overline{E}(\check{\boldsymbol{b}} - \overline{E}\check{\boldsymbol{b}})\,(\check{\boldsymbol{b}} - \overline{E}\check{\boldsymbol{b}})' = n^{-1}\boldsymbol{W} \otimes \boldsymbol{\Sigma}_{xx}^{-1}\ ,$$

$$plim\ \mathbf{S}(\check{\boldsymbol{b}}, \check{\boldsymbol{b}}) = \overline{\boldsymbol{V}}(\check{\boldsymbol{b}}, \check{\boldsymbol{b}})\ .$$

Note that the consistency of an estimator is insufficient to justify its use, since many estimators may be consistent. If the sample size grows to infinity, an additional criterion is necessary to enable a best estimator to be chosen from among the class of admissible estimators. If we confine ourselves to the class of consistent estimators we get the criterion of asymptotic efficiency. This is defined as follows. An estimator $\hat{\boldsymbol{\beta}}$ of $\boldsymbol{\beta}$ is called asymptotically efficient if it is consistent and if its asymptotic covariance matrix satisfies the condition

$$\overline{E}[(\tilde{\boldsymbol{\beta}} - \boldsymbol{\beta})\,(\tilde{\boldsymbol{\beta}} - \boldsymbol{\beta})'] - \overline{E}[(\hat{\boldsymbol{\beta}} - \boldsymbol{\beta})\,(\hat{\boldsymbol{\beta}} - \boldsymbol{\beta})'] \quad \text{is non-negative definite}$$

where $\tilde{\boldsymbol{\beta}}$ is any arbitrary consistent estimator.

It can be proved under (9.3.7) that $\check{\boldsymbol{b}}$ is asymptotically efficient within the class of consistent linear homogeneous estimators.

We now recall that the disturbance terms are contemporaneously uncorrelated i.e. $E[\boldsymbol{u}\boldsymbol{u}'] = \boldsymbol{W} \otimes \boldsymbol{T}$. If T^* is fixed then the prediction model corresponding to (9.3.7) is by analogy with (9.2.2)

$$\left.\begin{array}{lll} \boldsymbol{y}_* = \boldsymbol{Z}_*\boldsymbol{\beta} + \boldsymbol{u}_*\ , & \boldsymbol{Z}_* \text{ nonstochastic} & \\ E\boldsymbol{u}_* = \boldsymbol{0}\ , & E\boldsymbol{u}_*\boldsymbol{u}_*' = \boldsymbol{W}\ , & E\boldsymbol{u}\boldsymbol{u}_*' = \boldsymbol{0} \end{array}\right\} \tag{9.3.13}$$

We estimate $\boldsymbol{y}_*$ using classical prediction (cp. Theorem 9.3). This gives

$$\check{\boldsymbol{p}}_3 = \boldsymbol{Z}_*\check{\boldsymbol{b}}$$

with

$$plim\ \check{\boldsymbol{p}}_3 = \boldsymbol{Z}_*\ plim\ \check{\boldsymbol{b}} = \boldsymbol{Z}_*\boldsymbol{\beta} = \overline{E}\check{\boldsymbol{p}}_3$$

(see (9.3.9)). We may define the asymptotic risk function of a predictor $\boldsymbol{p}$ as follows

$$\overline{R}(\boldsymbol{p}) = \overline{E}(\boldsymbol{p} - \boldsymbol{y}_*)'\,(\boldsymbol{p} - \boldsymbol{y}_*) = n^{-1} \lim_{n\to\infty} E[\sqrt{n}(\boldsymbol{p} - \boldsymbol{y}_*)]'\,[\sqrt{n}\,(\boldsymbol{p} - \boldsymbol{y}_*)]\ . \tag{9.3.14}$$

For $\check{\boldsymbol{p}}_3$ this gives

$$\overline{R}(\check{\boldsymbol{p}}_3) = tr\,\boldsymbol{Z}_*[\overline{E}(\check{\boldsymbol{b}} - \boldsymbol{\beta})\,(\check{\boldsymbol{b}} - \boldsymbol{\beta})']\,\boldsymbol{Z}_*' + tr\,\overline{E}\boldsymbol{u}_*\boldsymbol{u}_*'$$
$$- 2\,tr\,\boldsymbol{Z}_*\overline{E}\{(\check{\boldsymbol{b}} - \boldsymbol{\beta})\,\boldsymbol{u}_*'\}\ .$$

Therefore using Theorem 9.4 along with assumptions (9.3.13) we have

$$\overline{R}(\check{\boldsymbol{p}}_3) = tr\,\boldsymbol{W} + tr\,\boldsymbol{Z}_*(n^{-1}\boldsymbol{W} \otimes \boldsymbol{\Sigma}_{xx}^{-1})\,\boldsymbol{Z}_*'\ . \tag{9.3.15}$$

This risk can be estimated consistently by

$$\hat{\overline{R}}(\breve{\boldsymbol{p}}_3) = tr\, \overline{\boldsymbol{W}} + tr\, \boldsymbol{Z}_*(n^{-1}\overline{\boldsymbol{W}} \otimes (\boldsymbol{X}'\boldsymbol{X})^{-1})\, \boldsymbol{Z}'_* \,. \tag{9.3.16}$$

Hence we have the following result.

Theorem 9.5 In the multivariate (classical) contemporaneously uncorrelated regression model the predictor

$$\breve{\boldsymbol{p}}_3 = \boldsymbol{Z}_*\breve{\boldsymbol{b}}$$

with

$$\overline{R}(\breve{\boldsymbol{p}}_3) = tr\, \boldsymbol{W} + tr\, \boldsymbol{Z}_*\overline{\boldsymbol{V}}(\boldsymbol{b}, \boldsymbol{b})\, \boldsymbol{Z}'_*$$

is consistent, asymptotically unbiased and $\overline{R}$-optimal in the class of homogeneous consistent linear predictors.

9.4 Prediction and the reduced form

The reduced form of the econometric model is

$$\boldsymbol{Y} = \boldsymbol{X\Pi} + \boldsymbol{V} \,. \tag{9.4.1}$$

The m-th equation has the form

$$\boldsymbol{y}_m = \boldsymbol{X\pi}_m + \boldsymbol{v}_m \qquad (m = 1, \dots, M) \,. \tag{9.4.2}$$

If we assume the errors to be contemporaneously uncorrelated then the covariance matrix will be

$$E\boldsymbol{v}_m\boldsymbol{v}'_{m'} = \sigma_{mm'}\boldsymbol{I} \,. \tag{9.4.3}$$

Here $\sigma_{mm'}$ is the (m, m')-th element of the matrix

$$\boldsymbol{\Sigma}_{vv} = E[\boldsymbol{v}(t)\, \boldsymbol{v}'(t)] \,.$$

We may arrange the M equations of (9.4.2) one under the other. This expresses the model in a form which is equivalent to (9.4.1), namely

$$\begin{pmatrix} \boldsymbol{y}_1 \\ \vdots \\ \boldsymbol{y}_M \end{pmatrix} = \begin{pmatrix} \boldsymbol{X} & \cdots & \boldsymbol{0} \\ \vdots & & \\ \boldsymbol{0} & \cdots & \boldsymbol{X} \end{pmatrix} \begin{pmatrix} \boldsymbol{\pi}_1 \\ \vdots \\ \boldsymbol{\pi}_M \end{pmatrix} + \begin{pmatrix} \boldsymbol{v}_1 \\ \vdots \\ \boldsymbol{v}_M \end{pmatrix} .$$

This may be summarized as

$$\boldsymbol{y} = \boldsymbol{Z\pi} + \boldsymbol{v}$$

where $E\boldsymbol{vv}' = \boldsymbol{\Sigma}_{vv} \otimes \boldsymbol{I}$.

The assumption $plim\; n^{-1}\boldsymbol{X}'\boldsymbol{V} = \boldsymbol{0}$ implies that the processes $\{\boldsymbol{x}(t)\}$ and $\{\boldsymbol{u}(t)\}$ are contemporaneously uncorrelated and that

$$plim\; n^{-1}\boldsymbol{X}'\boldsymbol{v}_m = plim\; n^{-1} \sum_{t=1}^{n} x(t)\, v_{mt} = 0 \,.$$

Hence the reduced form model may be represented as follows:

$$\left.\begin{aligned} &\boldsymbol{y} = \boldsymbol{Z\pi} + \boldsymbol{v}\,, \quad E\boldsymbol{v} = \boldsymbol{0}\,, \quad E\boldsymbol{vv}' = \boldsymbol{\Sigma}_{vv} \otimes \boldsymbol{I}\,,\\ &\boldsymbol{\Sigma}_{vv} \text{ positive definite}, \; plim\; n^{-1}\boldsymbol{Z}'\boldsymbol{v} = \boldsymbol{0}\,,\\ &plim\; n^{-1}\begin{pmatrix} tr\,\boldsymbol{v}_1\boldsymbol{v}_1' & \cdots & tr\,\boldsymbol{v}_1\boldsymbol{v}_M' \\ \vdots & & \\ tr\,\boldsymbol{v}_M\boldsymbol{v}_1' & \cdots & tr\,\boldsymbol{v}_M\boldsymbol{v}_M' \end{pmatrix} = \boldsymbol{\Sigma}_{vv}\,,\\ &plim\; n^{-1}(\boldsymbol{Z}'\boldsymbol{Z}) = \boldsymbol{\Sigma}_{xx} \otimes \boldsymbol{I}\,. \end{aligned}\right\} \tag{9.4.4}$$

The corresponding prediction model is (see (9.3.13))

$$\boldsymbol{y}_* = \boldsymbol{Z}_*\boldsymbol{\pi} + \boldsymbol{v}_*\,, \quad E\boldsymbol{v}_* = \boldsymbol{0}\,, \quad E\boldsymbol{v}_*\boldsymbol{v}_*' = \boldsymbol{\Sigma}_{vv}\,, \quad E\boldsymbol{vv}_*' = \boldsymbol{0}\,. \tag{9.4.5}$$

Therefore the theorems on estimation and prediction in the multivariate classical contemporaneously uncorrelated regression model follow through. From Theorem 9.5 the consistent predictor of the vector $\boldsymbol{y}_*$ of the M endogeneous variables is

$$\check{\boldsymbol{p}}_3 = \boldsymbol{Z}_*\hat{\boldsymbol{\pi}}_3 \tag{9.4.6}$$

where

$$\hat{\boldsymbol{\pi}}_3 = (\boldsymbol{Z}'\boldsymbol{Z})^{-1}\boldsymbol{Z}'\boldsymbol{y}\,, \quad \overline{V}(\hat{\boldsymbol{\pi}}_3, \hat{\boldsymbol{\pi}}_3) = n^{-1}\boldsymbol{\Sigma}_{vv} \otimes \boldsymbol{\Sigma}_{xx}^{-1}\,,$$

$$\overline{R}(\check{\boldsymbol{p}}_3) = tr\,\boldsymbol{\Sigma}_{vv} + tr\,\boldsymbol{Z}_*\overline{\boldsymbol{V}}(\hat{\boldsymbol{\pi}}_3, \hat{\boldsymbol{\pi}}_3)\boldsymbol{Z}_*'\,.$$

$\check{\boldsymbol{p}}_3$ is $\overline{R}$-optimal if no other information exists apart from that contained in (9.4.4).

9.5 The structural form and identifiability

In prediction the question of identifiability plays a different role from that which it plays in estimation problems. It affects the reliability of the predictions, but does not constrain the possibility of prediction itself. If we use the prior information necessary for the identification of structural parameters, then prediction is usually improved by using structural estimation procedures like two-stage least squares rather than reduced-form procedures. The parameters of reduced form and structural form are connected by the following equation

$$\boldsymbol{\Pi} = -\boldsymbol{D\Gamma}^{-1}\,.$$

If $\hat{\boldsymbol{D}}$ and $\hat{\boldsymbol{\Gamma}}$ are consistent estimators (see e.g. GOLDBERGER, 1964), then $\boldsymbol{\Pi}$ can be consistently estimated by

$$\hat{\boldsymbol{\Pi}} = -\hat{\boldsymbol{D}}\hat{\boldsymbol{\Gamma}}^{-1}\,.$$

Therefore the linear predictor

$$\hat{\boldsymbol{p}} = \boldsymbol{Z}_* \hat{\boldsymbol{\Pi}}$$

is consistent also, and

$$\bar{R}(\hat{\boldsymbol{p}}) = tr\, \boldsymbol{\Sigma}_{vv} + tr\, \boldsymbol{Z}_* \bar{\boldsymbol{V}}(\hat{\boldsymbol{\Pi}}, \hat{\boldsymbol{\Pi}})\, \boldsymbol{Z}'_* \,,$$

where $\bar{R}$ was defined at (9.3.14). Assertions concerning the $\bar{R}$-optimality of reduced form estimators based on structural form procedures cannot be given in general. This is because for instance the minimum variance property of a structural estimator does not carry over to the corresponding reduced form estimator. But there have been investigations concerning efficiency (KADIYALA (1970) investigated k-class estimators) and asymptotic efficiency (WICKENS (1969) and AMEMIYA (1966) investigated simultaneous equations with autocorrelated errors). In general structural estimators are asymptotically more efficient than reduced form estimators. Analogous assertions can presumably be made for the corresponding predictors.

9.6 Summary

This chapter generalizes previous results. Section 9.1 shows that the introduction of stochastic regressors need not affect the relative desirability of different predictors. Sections 9.2 and 9.3 investigate multivariate regression, while Section 9.4 and 9.5 consider various econometric formulations.

10

Minimax linear estimation

10.1 Ridge regression

The method of ridge regression suggested by HOERL and KENNARD (1970a, 1970b) will first be described. The idea is as follows. The risk $R(\hat{\boldsymbol{\beta}}) = \sigma^2 \, tr\, \boldsymbol{A}(\boldsymbol{X}'\boldsymbol{X})^{-1}$ of the OLSE $\hat{\boldsymbol{\beta}} = (\boldsymbol{X}'\boldsymbol{X})^{-1}\boldsymbol{X}'\boldsymbol{y}$ depends directly on the matrix $(\boldsymbol{X}'\boldsymbol{X})$. This risk increases as $\boldsymbol{X}'\boldsymbol{X}$ moves away from the form of an identity matrix. This led HOERL and KENNARD to propose the '*ridge*' estimator

$$\hat{\boldsymbol{\beta}}^* = [\boldsymbol{X}'\boldsymbol{X} + k\boldsymbol{I}]^{-1}\boldsymbol{X}'\boldsymbol{y} \tag{10.1.1}$$

where k is a given positive constant. We have already met this estimator in (1.3.3).

The question arises whether the relation

$$R(\hat{\boldsymbol{\beta}}) - R(\hat{\boldsymbol{\beta}}^*) > 0 \tag{10.1.2}$$

is valid or not. This was answered by HOERL and KENNARD in the following theorem.

Theorem 10.1 (HOERL and KENNARD, 1970a)
If

$$0 < k < \frac{\sigma^2}{\alpha_0^2} \tag{10.1.3}$$

where

$$\alpha_0^2 = \max_{i=1,\ldots,p} \alpha_i^2 \, ,$$

and

$$\boldsymbol{\alpha} = \boldsymbol{P}\boldsymbol{\beta} = (\alpha_1, \ldots, \alpha_P)'$$

and $\boldsymbol{P}$ is the orthogonal matrix such that $\boldsymbol{X}'\boldsymbol{X} = \boldsymbol{P}'\Lambda\boldsymbol{P}$ where Λ is the matrix of the eigenvalues of $\boldsymbol{X}'\boldsymbol{X}$, then the ridge estimator $\hat{\boldsymbol{\beta}}^*$ defined in (10.1.1) is better than the OLSE.

Alternatively in the terminology of Chapter 2, we may say that (10.1.3) defines an '*improvement region*' within which $\hat{\boldsymbol{\beta}}^*$ is better than $\hat{\boldsymbol{\beta}}$.

In the general linear model with correlated errors, i.e.

$$y = X\beta\,, \qquad u \sim (0, \sigma^2 W) \tag{10.1.4}$$

this theorem may be generalized by

$$\hat{\beta}(c) = [S + cI]^{-1} X'W^{-1}y = D_I^{-1}X'W^{-1}y \tag{10.1.5}$$

where $c \geqq 0$. (An alternative generalization would put W in place of I in (10.1.5), but that will not be considered here.) From (10.1.5) we deduce that

$$\textit{bias}\ \hat{\beta}(c) = -\,cD_I^{-1}\beta\,, \tag{10.1.6}$$

$$\textit{cov}\ \hat{\beta}(c) = V_{\hat{\beta}(c)} = \sigma^2 D_I^{-1}SD_I^{-1} \tag{10.1.7}$$

and

$$R[\hat{\beta}(c)] = \textit{tr}\ AD_I^{-1}(\sigma^2 S + c^2\beta\beta')\,D_I^{-1}\,. \tag{10.1.8}$$

This last formula may be rewritten in the form

$$R[\hat{\beta}(c)] = \gamma_1(c) + \gamma_2(c)$$

where

$$\gamma_1(c) = \sigma^2\ \textit{tr}\ AD_I^{-1}SD_I^{-1}$$

and

$$\gamma_2(c) = c^2\ \textit{tr}\ AD_I^{-1}\beta\beta' D_I^{-1}\,.$$

The following properties are easily proved:

1. the function $\gamma_1(c)$ is a continuous and monotonically decreasing function if $c \geqq 0$; (10.1.9)
2. the function $\gamma_2(c)$ is a continuous and monotonically increasing function if $c \geqq 0$; 10.1.10)
3. in a region around zero, $\gamma_1(.)$ decreases more rapidly for increasing c than $\gamma_2(.)$ increases.

As $\gamma_2(c)$ depends on the unknown β, only qualitative interpretations can be made (see Figure 10.1.1). However, we can prove the following result.

Theorem 10.2 If the condition

$$c \leqq c_0 = \frac{2\sigma^2}{\beta'\beta} \tag{10.1.11}$$

is fulfilled then the ridge estimator defined in (10.1.5) is better than the GLSE, i.e.

$$R(b) - R[\hat{\beta}(c)] \geqq 0 \quad \text{if} \quad 0 < c \leqq c_0\,. \tag{10.1.12}$$

Proof From $R(b) = \sigma^2\ \textit{tr}\ AS^{-1}$ and (10.1.8) we deduce that

$$\begin{aligned} R(b) - R[\hat{\beta}(c)] &= \textit{tr}\ A\{\sigma^2 S^{-1} - D_I^{-1}(\sigma^2 S + c^2\beta\beta')\,D_I^{-1}\} \\ &= \textit{tr}\ AB\,, \quad \text{say}\,. \end{aligned}$$

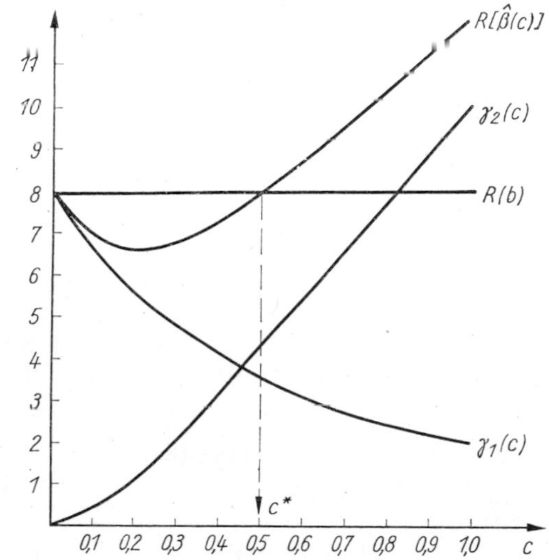

Figure 10.1.1 Interrelationships between the functions $\gamma_1(c)$, $\gamma_2(c)$, $R[\hat{\beta}(c)]$ and $R(\boldsymbol{b})$. The figure shows that $\hat{\beta}(c)$ is better than the GLSE $\boldsymbol{b}$ if c is less than a certain c^* which in general is unknown.

Therefore, to satisfy (10.1.12) the matrix $\boldsymbol{B}$ must be nonnegative definite. But

$$\begin{aligned}\boldsymbol{B} &= \boldsymbol{D}_I^{-1}(\sigma^2\boldsymbol{D}_I\boldsymbol{S}^{-1}\boldsymbol{D}_I - \sigma^2\boldsymbol{S} - c^2\boldsymbol{\beta}\boldsymbol{\beta}')\,\boldsymbol{D}_I^{-1}\\ &= \boldsymbol{D}_I^{-1}\tilde{\boldsymbol{B}}\boldsymbol{D}_I^{-1}\,, \quad \text{say}\,.\end{aligned}$$

As $\boldsymbol{D}_I$ is positive definite an equivalent condition is that $\tilde{\boldsymbol{B}}$ should be nonnegative definite, where

$$\begin{aligned}\tilde{\boldsymbol{B}} &= \sigma^2(\boldsymbol{S} + c\boldsymbol{I})\,\boldsymbol{S}^{-1}(c\boldsymbol{I} + \boldsymbol{S}) - \sigma^2\boldsymbol{S} - c^2\boldsymbol{\beta}\boldsymbol{\beta}'\\ &= \sigma^2c^2\boldsymbol{S}^{-1} + [2\sigma^2c\boldsymbol{I} - c^2\boldsymbol{\beta}\boldsymbol{\beta}']\,.\end{aligned}$$

$\tilde{\boldsymbol{B}}$ is positive definite if

$$[2\sigma^2\boldsymbol{I} - c^2\boldsymbol{\beta}\boldsymbol{\beta}'] \tag{10.1.13}$$

is nonnegative definite. Now the eigenvalues of this matrix are $2\sigma^2$ and $2\sigma^2 - c^2\boldsymbol{\beta}'\boldsymbol{\beta}$. As is well known, the eigenvalues of a nonnegative definite matrix must be nonnegative. Therefore the theorem follows.

It follows from Theorem 10.2 that prior knowledge of a region

$$\mathfrak{B} = \{\boldsymbol{\beta} \in E^p : \boldsymbol{\beta}'\boldsymbol{\beta} \leqq \sigma^2k\} \tag{10.1.14}$$

would allow one to deduce a ridge estimator

$$\hat{\boldsymbol{\beta}}(c) = (\boldsymbol{S} + c\boldsymbol{I})^{-1}\,\boldsymbol{X}'\boldsymbol{W}^{-1}\boldsymbol{y}$$

which is uniformly better than the GLSE.

The next question to arise is which value of c is optimal in the sense that $R[\hat{\beta}(c)]$ is minimized.

An answer is given in Section 10.2, where it is also shown that ridge regression is a special case of the more general method of minimax linear estimation which is based on the prior knowledge of regions such as that given in (10.1.14).

10.2 The minimax principle

10.2.1 Introduction. We turn now to consider the notion of minimax linear estimation (MMLE) as outlined by Kuks and Olman (1971, 1972) and Kuks (1972). Using the by now familiar linear model

$$\boldsymbol{y} = \boldsymbol{X\beta} + \boldsymbol{u}\,, \qquad \boldsymbol{u} \sim (\boldsymbol{0}, \sigma^2 \boldsymbol{W}) \tag{10.2.1}$$

and the quadratic loss function

$$R(\hat{\boldsymbol{\beta}}, \boldsymbol{\beta}, \boldsymbol{A}) = R(\hat{\boldsymbol{\beta}}) = E(\hat{\boldsymbol{\beta}} - \boldsymbol{\beta})' \boldsymbol{A}(\hat{\boldsymbol{\beta}} - \boldsymbol{\beta}) \tag{10.2.2}$$

the minimax principle states that one should use the estimator $\boldsymbol{b}_0$ that satisfies

$$R(\boldsymbol{b}_0) = \min_{\boldsymbol{b}} \sup_{\boldsymbol{\beta} \in \mathfrak{B}} R(\boldsymbol{b}, \boldsymbol{\beta}, \boldsymbol{A})\,. \tag{10.2.3}$$

Here we generally constrain $\boldsymbol{b}_0$ to be a linear estimator, of the form

$$\boldsymbol{b}_0 = \boldsymbol{C}'\boldsymbol{y} + \boldsymbol{d} \qquad \text{(heterogeneous)}\,, \tag{10.2.4}$$

or

$$\boldsymbol{b}_0 = \boldsymbol{C}'\boldsymbol{y} \qquad \text{(homogeneous)}\,, \tag{10.2.5}$$

and $\boldsymbol{\beta}$ is also constrained to lie in the set

$$\mathfrak{B} = \{\boldsymbol{\beta}\colon \boldsymbol{\beta}'\boldsymbol{T\beta} \leqq k\} \tag{10.2.6}$$

where $k \geqq 0$ is a given constant term and $\boldsymbol{T}$ is a known positive definite matrix. In other words, we assume that some known constraint exists on the length of $\boldsymbol{\beta}$, and this is expressed by the inequality in (10.2.6).

Now from (10.2.2) and (10.2.4) we can show that

$$\begin{aligned} R(\boldsymbol{b}_0, \boldsymbol{\beta}, \boldsymbol{A}) &= \sigma^2 \operatorname{tr} \boldsymbol{AC}'\boldsymbol{WC} + \boldsymbol{\beta}'(\boldsymbol{C}'\boldsymbol{X} - \boldsymbol{I})' \boldsymbol{A}(\boldsymbol{C}'\boldsymbol{X} - \boldsymbol{I})\, \boldsymbol{\beta} \\ &\quad + \boldsymbol{d}'\boldsymbol{Ad} + 2\boldsymbol{d}'\boldsymbol{A}(\boldsymbol{C}'\boldsymbol{X} - \boldsymbol{I})\, \boldsymbol{\beta}\,. \end{aligned} \tag{10.2.7}$$

When $\boldsymbol{b}_0$ is homogeneous, so that $\boldsymbol{d} = \boldsymbol{0}$ as in (10.2.5), then the last two terms in (10.2.7) fall away.

Writing $\boldsymbol{d} = \boldsymbol{0}$ in (10.2.7) we have for homogeneous estimators

$$R(\boldsymbol{b}_0, \boldsymbol{\beta}, \boldsymbol{A}) = \sigma^2 \operatorname{tr} \boldsymbol{AC}'\boldsymbol{WC} + \boldsymbol{\beta}'\boldsymbol{T}^{1/2}\tilde{\boldsymbol{A}}\boldsymbol{T}^{1/2}\boldsymbol{\beta} \tag{10.2.8}$$

where

$$\tilde{\boldsymbol{A}} = \boldsymbol{T}^{-1/2}(\boldsymbol{C}'\boldsymbol{X} - \boldsymbol{I})' \boldsymbol{A}(\boldsymbol{C}'\boldsymbol{X} - \boldsymbol{I})\, \boldsymbol{T}^{-1/2}\,. \tag{10.2.9}$$

Now

$$\min_{\beta} \frac{\beta' T^{1/2} \tilde{A} T^{1/2} \beta}{\beta' T \beta} = \lambda_1(\tilde{A}) \,,$$

using the extremal property of λ_1, the largest eigenvalue. Therefore from (10.2.6)

$$\sup_{\beta \in \mathfrak{B}} \beta' T^{1/2} \tilde{A} T^{1/2} \beta = k \lambda_1(\tilde{A}) \,, \tag{10.2.10}$$

and from (10.2.8)

$$\sup_{\beta \in \mathfrak{B}} R(b_0, \beta, A) = \sigma^2 \, tr \, AC'WC + k \lambda_1(\tilde{A}) \,. \tag{10.2.11}$$

The minimax linear estimator (MMLE) is obtained by finding the value of C which minimizes (10.2.11). However, since $\tilde{A}$ depends on C, this problem is complicated. It was examined by LÄUTER (1975), using various equivalence theorems from games theory. However, we shall concentrate now on the more tractable special case where A has rank one, so that $A = aa'$. In this case $\tilde{A}$ from (10.2.9) is given by

$$\tilde{A} = \tilde{a}\tilde{a}' \quad \text{where } \tilde{a} = A^{-1/2}(C'X - T)' \, a \,. \tag{10.2.12}$$

Now $\lambda_1(\tilde{A}) = \tilde{a}'\tilde{a}$, so using (10.2.11)

$$\begin{aligned} \sup_{\beta \in \mathfrak{B}} R(b_0, \beta, aa') &= \sigma^2 a'C'WCa + k a'(C'X - I) \, T^{-1}(C'X - I)' \, a \\ &= r(b_0) \,, \quad \text{say.} \end{aligned} \tag{10.2.13}$$

Using Theorem A 26 we deduce that

$$\frac{1}{2} \frac{\partial r(b_0)}{\partial C} = (\sigma^2 W + k X T^{-1} X') \, (aa' - k X T^{-1} aa') \,.$$

This is zero when

$$C' = \hat{C}_1' = k(\sigma^2 W + k X T^{-1} X')^{-1} \, X T^{-1} \tag{10.2.14}$$

or equivalently, putting $X'W^{-1}X = S$ (see KUKS and OLMAN, 1972),

$$\hat{C}_1' = (k^{-1}\sigma^2 T + S)^{-1} \, X' W^{-1} \,. \tag{10.2.15}$$

This yields a biased estimator whose properties are described in the following theorem.

Theorem 10.3 In the model (10.2.1), with constraint (10.2.6), the minimax linear estimator with $A = aa'$ in (10.2.2) is

$$\hat{\beta}_1 = \hat{C}_1' y = (k^{-1}\sigma^2 T + S)^{-1} \, X' W^{-1} y \,. \tag{10.2.16}$$

This estimator has

$$bias \, (\hat{\beta}_1) = -k^{-1}\sigma^2 D^{-1} T \beta \,,$$

$$variance \, (\hat{\beta}_1) = \sigma^2 D^{-1} S D^{-1} = V(\hat{\beta}_1) \,, \quad \text{say,}$$

and risk

$$r(\hat{\boldsymbol{\beta}}_1) = \sigma^2 \boldsymbol{a}'\boldsymbol{D}^{-1}\boldsymbol{a}\ , \tag{10.2.17}$$

where

$$\boldsymbol{D} = k^{-1}\sigma^2 \boldsymbol{T} + \boldsymbol{S}\ .$$

It follows that the MMLE given by (10.2.16) is a generalization of the ridge estimator defined in (10.1.5). Alternatively, putting $\boldsymbol{T} = \boldsymbol{I}$ and $c = k^{-1}\sigma^2$, (10.1.5) is a special case of (10.2.16). *That is, minimax linear estimation is a more general method than ridge regression.*

For convenience we evaluate the R-risk of the minimax estimator $\hat{\boldsymbol{\beta}}_1$. (Note that this is different from $\hat{\boldsymbol{\beta}}_1 = \boldsymbol{\beta}$ given in (4.2.23).) Now

$$\begin{aligned} R(\hat{\boldsymbol{\beta}}_1) &= tr\, \boldsymbol{A}\{\boldsymbol{V}(\hat{\boldsymbol{\beta}}_1) + (bias\ \hat{\boldsymbol{\beta}}_1)\ (bias\ \hat{\boldsymbol{\beta}}_1)'\} \\ &= \sigma^2\ tr\, \boldsymbol{A}\boldsymbol{D}^{-1}(\boldsymbol{S} + k^{-2}\sigma^2 \boldsymbol{T}\boldsymbol{\beta}\boldsymbol{\beta}'\boldsymbol{T})\, \boldsymbol{D}^{-1}\ . \end{aligned} \tag{10.2.18}$$

Alternatively we may set $\boldsymbol{C} = \hat{\boldsymbol{C}}_1$ and $\boldsymbol{d} = \boldsymbol{0}$ in the general risk function (10.2.7), which gives

$$\begin{aligned} R(\hat{\boldsymbol{\beta}}_1) &= \sigma^2\ tr\, \boldsymbol{A}\hat{\boldsymbol{C}}_1'\boldsymbol{W}\hat{\boldsymbol{C}}_1 + tr\, \boldsymbol{A}(\hat{\boldsymbol{C}}_1'\boldsymbol{X} - \boldsymbol{I})\ \boldsymbol{\beta}\boldsymbol{\beta}'(\hat{\boldsymbol{C}}_1'\boldsymbol{X} - \boldsymbol{I})' \\ &= tr\, \boldsymbol{A}(\boldsymbol{D}^{-1}\boldsymbol{X}'\boldsymbol{W}^{-1}\boldsymbol{P}\boldsymbol{W}^{-1}\boldsymbol{X}\boldsymbol{D}^{-1} + \boldsymbol{\beta}\boldsymbol{\beta}' - 2\boldsymbol{D}^{-1}\boldsymbol{S}\boldsymbol{\beta}\boldsymbol{\beta}')\ , \end{aligned} \tag{10.2.19}$$

where $\boldsymbol{P} = (\boldsymbol{X}\boldsymbol{\beta}\boldsymbol{\beta}'\boldsymbol{X}' + \sigma^2\boldsymbol{W})$.

From the unbiasedness condition $\boldsymbol{C}'\boldsymbol{X} - \boldsymbol{I} = \boldsymbol{0}$ and (10.2.11) it follows that

$$\sup_{\boldsymbol{\beta}\in\mathfrak{B}} R(\boldsymbol{C}'\boldsymbol{y}, \boldsymbol{\beta}, \boldsymbol{A}) = R(\boldsymbol{C}'\boldsymbol{y}, \boldsymbol{\beta}, \boldsymbol{A})\ .$$

Hence we have the following result.

Theorem 10.4. In the case of homogeneous unbiased estimation the MMLE and the R-optimal estimator coincide.

10.2.2 Comparison with *R*-optimal estimators. Using Theorem 10.4 we have $R(\boldsymbol{b}) = r(\boldsymbol{b})$ and therefore using (10.2.17) and $R(\boldsymbol{b}) = \sigma^2\ tr\, \boldsymbol{A}\boldsymbol{S}^{-1}$ it follows for $\boldsymbol{A} = \boldsymbol{a}\boldsymbol{a}'$ that

$$r(\boldsymbol{b}) - r(\hat{\boldsymbol{\beta}}_1) = \sigma^2\boldsymbol{a}'\{\boldsymbol{S}^{-1} - (k^{-1}\sigma^2\boldsymbol{T} + \boldsymbol{S})^{-1}\}\ \boldsymbol{a} \geqq 0\ . \tag{10.2.20}$$

To compare the R-risk of the two biased estimators $\hat{\boldsymbol{\beta}}_1$ and $\hat{\boldsymbol{\beta}}_2$ given in (10.2.16) and (4.2.10) we use (10.2.19) and (4.2.24) to derive the following:

$$\begin{aligned} R(\hat{\boldsymbol{\beta}}_1) - R(\hat{\boldsymbol{\beta}}_2) &= tr\, \boldsymbol{A}[\boldsymbol{D}^{-1}\boldsymbol{X}'\boldsymbol{W}^{-1}\boldsymbol{P}\boldsymbol{W}^{-1}\boldsymbol{X}\boldsymbol{D}^{-1} \\ &\qquad - 2\boldsymbol{D}^{-1}\boldsymbol{X}'\boldsymbol{W}^{-1}\boldsymbol{X}\boldsymbol{\beta}\boldsymbol{\beta}' + \boldsymbol{\beta}\boldsymbol{\beta}'\boldsymbol{X}'\boldsymbol{P}^{-1}\boldsymbol{X}\boldsymbol{\beta}\boldsymbol{\beta}'] \\ &= tr\, \boldsymbol{A}[(\boldsymbol{D}^{-1}\boldsymbol{X}'\boldsymbol{W}^{-1} - \boldsymbol{\beta}\boldsymbol{\beta}'\boldsymbol{X}'\boldsymbol{P}^{-1})\, \boldsymbol{P}(\boldsymbol{W}^{-1}\boldsymbol{X}\boldsymbol{D}^{-1} \\ &\qquad - \boldsymbol{P}^{-1}\boldsymbol{X}\boldsymbol{\beta}\boldsymbol{\beta}')] \geqq 0\ . \end{aligned} \tag{10.2.21}$$

This expression is nonnegative because the matrix in square brackets is nonnegative definite.

If there is no restriction on $\boldsymbol{\beta}$, in other words if $\mathfrak{B} = E^P$ or equivalently if $k = \infty$, then we have $\lim_{k\to\infty} \boldsymbol{D} = \boldsymbol{S}$ and $\lim_{k\to\infty} \hat{\boldsymbol{\beta}}_1 = \boldsymbol{b}$. Therefore MIMSEE is a special case of the MMLE. In this case (10.2.21) expresses the fact that $\hat{\boldsymbol{\beta}}_2$ has the property of R-optimality in the class of all homogeneous estimators (see also (4.2.22) where this result could be deduced using the fact that any restriction can only increase the risk function.)

Taking the limiting case of (10.2.21) we see that

$$R(\boldsymbol{b}) - R(\hat{\boldsymbol{\beta}}_2) \geqq 0 . \tag{10.2.22}$$

Comparing $\hat{\boldsymbol{\beta}}_1$ and $\hat{\boldsymbol{\beta}}_2$ according to r-risk, we know immediately from Theorem 10.3 that $r(\hat{\boldsymbol{\beta}}_1) < r(\hat{\boldsymbol{\beta}}_2)$ must hold. Nevertheless we examine the $r(\hat{\boldsymbol{\beta}}_2)$-risk. Bibby (1972) gives $\hat{\boldsymbol{\beta}}_2$ from (4.2.10) in the following equivalent form

$$\hat{\boldsymbol{\beta}}_2 = (\sigma^2 + \boldsymbol{\beta}'\boldsymbol{S}\boldsymbol{\beta})^{-1}\, \boldsymbol{\beta}\boldsymbol{\beta}'\boldsymbol{X}'\boldsymbol{W}^{-1}\boldsymbol{y} , \tag{10.2.23}$$

which has risk

$$R(\hat{\boldsymbol{\beta}}_2) = \sigma^2(\sigma^2 + \boldsymbol{\beta}'\boldsymbol{S}\boldsymbol{\beta})^{-1}\, \boldsymbol{a}'\boldsymbol{\beta}\boldsymbol{\beta}'\boldsymbol{a} . \tag{10.2.24}$$

We may evaluate

$$\sup_{\boldsymbol{\beta}\in\mathfrak{B}} \boldsymbol{\beta}'\boldsymbol{a}\boldsymbol{a}'\boldsymbol{\beta} = k\boldsymbol{a}'\boldsymbol{T}^{-1}\boldsymbol{a} ,$$

$$\sup_{\boldsymbol{\beta}\in\mathfrak{B}} \boldsymbol{\beta}'\boldsymbol{S}\boldsymbol{\beta} = k\lambda_1(\boldsymbol{T}^{-1/2}\boldsymbol{S}\boldsymbol{T}^{-1/2}) ,$$

and

$$r(\hat{\boldsymbol{\beta}}_2) = \sup_{\boldsymbol{\beta}\in\mathfrak{B}} R(\hat{\boldsymbol{\beta}}_2) \geqq \sigma^2 \sup_{\boldsymbol{\beta}\in\mathfrak{B}} \boldsymbol{\beta}'\boldsymbol{a}\boldsymbol{a}'\boldsymbol{\beta}\, [\sup_{\boldsymbol{\beta}\in\mathfrak{B}} (\sigma^2 + \boldsymbol{\beta}'\boldsymbol{S}\boldsymbol{\beta})]^{-1}$$

$$= \frac{\sigma^2 k\boldsymbol{a}'\boldsymbol{T}^{-1}\boldsymbol{a}}{\sigma^2 + k\lambda_1(\boldsymbol{T}^{-1/2}\boldsymbol{S}\boldsymbol{T}^{-1/2})} = k^*\boldsymbol{a}'\boldsymbol{T}^{-1}\boldsymbol{a}$$

with $k^{*-1} = k^{-1} + \sigma^{-2}\lambda_1(\boldsymbol{T}^{-1/2}\boldsymbol{S}\boldsymbol{T}^{-1/2})$. Therefore $k^*\boldsymbol{a}'\boldsymbol{T}^{-1}\boldsymbol{a}$ is a lower bound for $r(\hat{\boldsymbol{\beta}}_2)$. The question arises whether or not it is also an upper bound for $r(\hat{\boldsymbol{\beta}}_1)$. Now the relation

$$k^*\boldsymbol{a}'\boldsymbol{T}^{-1}\boldsymbol{a} \geqq \boldsymbol{a}'(k^{-1}\boldsymbol{T} + \sigma^{-2}\boldsymbol{S})^{-1}\, \boldsymbol{a} = r(\hat{\boldsymbol{\beta}}_1)$$

holds whenever

$$(k^{-1} - k^{*-1})\, \boldsymbol{T} + \sigma^{-2}\boldsymbol{S} = \sigma^2\boldsymbol{S} - \boldsymbol{T}\lambda_1(\boldsymbol{T}^{-1/2}\boldsymbol{S}\boldsymbol{T}^{-1/2})$$

is a nonnegative definite matrix. This is true for example when $\boldsymbol{T} = \boldsymbol{S}$ and $\sigma^2 < 1$.

10.3 Optimal substitution of σ^2

Unfortunately the minimax linear estimator (MMLE) $\hat{\boldsymbol{\beta}}_1$ given by (10.2.16) contains the unknown parameter σ and therefore it is not practicable. So we have to use sample or prior information to substitute an estimator for σ. One could propose the substitution by s, the well known estimator given in (4.2.25). But s is stochastic, so that the calculation of the risk function would be very difficult (see THEIL 1963 for approximations to such a risk function). Therefore we confine ourselves to the substitution of σ^2 by nonstochastic values $c > 0$. This gives the estimator

$$\hat{\boldsymbol{\beta}}_c = (k^{-1}c\boldsymbol{T} + \boldsymbol{S})^{-1}\boldsymbol{X}'\boldsymbol{W}^{-1}\boldsymbol{y} = \boldsymbol{D}_c^{-1}\boldsymbol{X}'\boldsymbol{W}^{-1}\boldsymbol{y} \tag{10.3.1}$$

which has

$$bias\ (\hat{\boldsymbol{\beta}}_c) = -k^{-1}c\boldsymbol{D}_c^{-1}\boldsymbol{T}\boldsymbol{\beta}\ , \tag{10.3.2}$$

$$variance\ (\hat{\boldsymbol{\beta}}_c) = \sigma^2\boldsymbol{D}_c^{-1}\boldsymbol{S}\boldsymbol{D}_c^{-1} = \boldsymbol{V}(\hat{\boldsymbol{\beta}}_c)\ , \quad \text{say,} \tag{10.3.3}$$

and

$$\sup_{\boldsymbol{\beta}\in\mathfrak{B}} \boldsymbol{a}'\ (bias\ \hat{\boldsymbol{\beta}}_c)\ (bias\ \hat{\boldsymbol{\beta}}_c)'\ \boldsymbol{a} = k^{-1}c^2\boldsymbol{a}'\boldsymbol{D}_c^{-1}\boldsymbol{T}\boldsymbol{D}_c^{-1}\boldsymbol{a}\ . \tag{10.3.4}$$

Therefore the risk is

$$r(\hat{\boldsymbol{\beta}}_c) = \sigma^2\boldsymbol{a}'\boldsymbol{D}_c^{-1}(\boldsymbol{S} + k^{-1}c^2\sigma^{-2}\boldsymbol{T})\,\boldsymbol{D}_c^{-1}\boldsymbol{a}\ . \tag{10.3.5}$$

The relation

$$r(\hat{\boldsymbol{\beta}}_1) \leqq r(\hat{\boldsymbol{\beta}}_c)$$

follows immediately from Theorem 10.3. We first seek values of c such that $\hat{\boldsymbol{\beta}}_c$ is r-better than the GLSE $\boldsymbol{b}$ and then seek to maximize the difference between the two risks, i.e. such that

$$\max_c \{r(\boldsymbol{b}) - r(\hat{\boldsymbol{\beta}}_c)\} = r(\boldsymbol{b}) - r(\hat{\boldsymbol{\beta}}_{c*}) \geqq 0\ .$$

Using (10.3.5), $r(\boldsymbol{b}) = \sigma^2\, tr\, \boldsymbol{A}\boldsymbol{S}^{-1}$, and Theorem 10.4, we get

$$r(\boldsymbol{b}) - r(\hat{\boldsymbol{\beta}}_c) = \sigma^2\boldsymbol{a}'[\boldsymbol{S}^{-1} - \boldsymbol{D}_c^{-1}(\boldsymbol{S} + k^{-1}c^2\sigma^{-2}\boldsymbol{T})\,\boldsymbol{D}_c^{-1}]\,\boldsymbol{a}\ . \tag{10.3.6}$$

The matrix in square brackets is nonnegative definite iff

$$\boldsymbol{D}_c\boldsymbol{S}^{-1}\boldsymbol{D}_c - (\boldsymbol{S} + k^{-1}c^2\sigma^{-2}\boldsymbol{T}) = k^{-2}c^2\boldsymbol{T}\boldsymbol{S}^{-1}\boldsymbol{T} + k^{-1}c(2 - c\sigma^{-2})\,\boldsymbol{T}$$

is nonnegative definite. A sufficient condition for this is given by KUKS and OLMAN (1972, p. 71) as

$$c \leqq 2\sigma^2\ . \tag{10.3.7}$$

To maximize (10.3.6) let us put $\tilde{\boldsymbol{T}} = k^{-1}c\boldsymbol{T}$. Since $\tilde{\boldsymbol{T}}^{-1}\boldsymbol{D}_c\tilde{\boldsymbol{T}}^{-1} - \tilde{\boldsymbol{T}}^{-1}$ is nonnegative definite, the same is true for $\tilde{\boldsymbol{T}} - \tilde{\boldsymbol{T}}\boldsymbol{D}_c^{-1}\tilde{\boldsymbol{T}}$. Therefore

$$\frac{\partial}{\partial c}\, r(\boldsymbol{\beta}_c) = -2\sigma^2c^{-1}(1 - c\sigma^{-2})\,(\tilde{\boldsymbol{T}} - \tilde{\boldsymbol{T}}\boldsymbol{D}_c^{-1}\tilde{\boldsymbol{T}})\ .$$

Hence we deduce that $r(\hat{\boldsymbol{\beta}}_c)$ is minimal when $c = \sigma^2$ (in accordance with Theorem 10.3), and moreover

$$\begin{aligned} r(\hat{\boldsymbol{\beta}}_c) &\text{ is monotonically increasing for } c > \sigma^2, \\ r(\hat{\boldsymbol{\beta}}_c) &\text{ is monotonically decreasing for } c < \sigma^2. \end{aligned} \tag{10.3.8}$$

We now assume that prior knowledge gives an interval such that

$$0 < \sigma_1^2 < \sigma^2 < \sigma_2^2 < \infty . \tag{10.3.9}$$

The three conditions (10.3.7), (10.3.8) and (10.3.9) are combined optimally when $c^* = 2\sigma_1^2$. This gives

Theorem 10.5 Given an interval such that $\sigma_1^2 < \sigma^2 < \sigma_2^2$, then

$$\hat{\boldsymbol{\beta}}_{2\sigma_1^2} = (2\sigma_1^2 k\boldsymbol{T} + \boldsymbol{S})^{-1} \boldsymbol{X}'\boldsymbol{W}^{-1}\boldsymbol{y}$$

is r-better than the GLSE $\boldsymbol{b}$, and moreover is optimal in the sense that

$$\max_{c \leqq 2\sigma_1^2} \{r(\boldsymbol{b}) - r(\hat{\boldsymbol{\beta}}_c)\} = r(\boldsymbol{b}) - r(\hat{\boldsymbol{\beta}}_{2\sigma_1^2}) .$$

Using the prior information (10.3.9) therefore gives a practicable estimator.

An alternative way of overcoming the unknown σ^2 in the minimax linear estimator (10.2.16) is to use a restriction on $\boldsymbol{\beta}$ such as

$$\boldsymbol{\beta}'\boldsymbol{T}\boldsymbol{\beta} \leqq k\sigma^2 . \tag{10.3.10}$$

We see that with $\boldsymbol{T}^* = \boldsymbol{T}\sigma^{-2}$ this can also be written $\boldsymbol{\beta}'\boldsymbol{T}^*\boldsymbol{\beta} \leqq k$ as in (10.2.6). This leads to the following result.

Theorem 10.6 In the generalized regression model given by (10.2.1), the MMLE under the constraint (10.3.10) is

$$(k^{-1}\boldsymbol{T} + \boldsymbol{S})^{-1} \boldsymbol{X}'\boldsymbol{W}^{-1}\boldsymbol{y} . \tag{10.3.11}$$

It may well be that in practice (10.2.6), i.e. the restriction of $\boldsymbol{\beta}$ alone is more useful than the restriction in (10.3.10) of the "coefficient of variation" $\boldsymbol{\beta}/\sigma$.

10.4 A numerical example

We confine ourselves to a classical regression model with $p = 3$ regressors

$$y_i = X_{1i}\beta_1 + X_{2i}\beta_2 + X_{3i}\beta_3 + u_i ,$$

where

$$\boldsymbol{u} \sim (\boldsymbol{0}, \sigma^2\boldsymbol{I}) .$$

We examine the following sample of size six, taken from GOLDBERGER (1964, p. 160):

$$\boldsymbol{y} = \begin{pmatrix} 0 \\ 2 \\ 1 \\ 2 \\ -1 \\ 1 \end{pmatrix}, \qquad \boldsymbol{X} = \begin{pmatrix} 1 & -1 & 0 \\ 1 & 0 & 1 \\ 1 & 1 & 0 \\ 1 & 2 & 1 \\ 1 & 0 & -1 \\ 1 & 0 & 0 \end{pmatrix}.$$

This gives

$$\boldsymbol{X}'\boldsymbol{X} = \begin{pmatrix} 6 & 2 & 1 \\ 2 & 6 & 2 \\ 1 & 2 & 3 \end{pmatrix}, \qquad \boldsymbol{X}'\boldsymbol{y} = \begin{pmatrix} 5 \\ 5 \\ 5 \end{pmatrix}, \qquad |\boldsymbol{X}'\boldsymbol{X}| = 74\,,$$

and

$$(\boldsymbol{X}'\boldsymbol{X})^{-1} = 1/74 \begin{pmatrix} 14 & -4 & -2 \\ -4 & 17 & -10 \\ -2 & -10 & 32 \end{pmatrix}.$$

Therefore the OLSE is

$$\hat{\boldsymbol{\beta}} = 1/74 \begin{pmatrix} 40 \\ 15 \\ 100 \end{pmatrix} = \begin{pmatrix} 0.541 \\ 0.203 \\ 1.352 \end{pmatrix}.$$

The vector of OLS fitted values is

$$\hat{\boldsymbol{y}} = \boldsymbol{X}\hat{\boldsymbol{\beta}} = 1/74 \begin{pmatrix} 25 \\ 140 \\ 55 \\ 170 \\ -60 \\ 40 \end{pmatrix}, \quad \text{the } OLS \text{ residuals are } \hat{\boldsymbol{u}} = \boldsymbol{y} - \boldsymbol{X}\hat{\boldsymbol{\beta}} = 1/74 \begin{pmatrix} -25 \\ 8 \\ 19 \\ -22 \\ -14 \\ 34 \end{pmatrix}$$

and the sample variance is

$$s^2 = \hat{\boldsymbol{u}}'\hat{\boldsymbol{u}}(n-p)^{-1} = \frac{2886}{74^2 \cdot 3} = \frac{13}{74}\,.$$

If $\boldsymbol{u}$ is normally distributed, then $(n-p)\, s^2/\sigma^2$ has a χ^2_{n-p} distribution, and a one sided confidence interval is given by

$$P\left(\frac{s^2(n-p)}{\chi^2_\alpha} < \sigma^2\right) = \alpha\,.$$

If $\alpha = 0.05$ then the lower bound on σ^2 is 0.0674. Therefore we may choose $\sigma_1^2 = 0.05$, as a number below which the true value of σ^2 is unlikely to be.

Now, if we know a priori that the following relations hold:

$$\beta_1^2 \leqq 1 , \quad \beta_2^2 \leqq 1 , \quad \beta_3^2 \leqq 2 ,$$

then $\boldsymbol{\beta}'\boldsymbol{T}\boldsymbol{\beta} \leqq 1$ with

$$\boldsymbol{T} = \begin{pmatrix} 1 & 0 & 0 \\ 0 & 1 & 0 \\ 0 & 0 & 0.5 \end{pmatrix} .$$

Therefore using $\hat{\beta}_{2\sigma_1^2}$ from Theorem 10.5 with $\sigma_1^2 = 0.05$ and $k = 1$ we obtain

$$\hat{\boldsymbol{\beta}}_{2\sigma_1^2} = \hat{\boldsymbol{\beta}}_{0.1} = (0.1\,\boldsymbol{T} + \boldsymbol{X}'\boldsymbol{X})^{-1}\,\boldsymbol{X}'\boldsymbol{y} .$$

Now

$$(0.1\,\boldsymbol{T} + \boldsymbol{X}'\boldsymbol{X})^{-1} = (79.25)^{-1} \begin{pmatrix} 14.61 & -4.10 & -2.10 \\ -4.10 & 17.61 & -10.20 \\ -2.10 & -10.20 & 33.21 \end{pmatrix} .$$

Therefore

$$\hat{\boldsymbol{\beta}}_{0.1} = (79.25)^{-1} \begin{pmatrix} 42.55 \\ 16.55 \\ 104.55 \end{pmatrix} = \begin{pmatrix} 0.537 \\ 0.208 \\ 1.319 \end{pmatrix}$$

We can also calculate the two risk functions for example in the case when $\boldsymbol{a}' = (1, 1, 1)$. This gives

$$r(\hat{\boldsymbol{\beta}}) = \sigma^2\, 31/74 = \sigma^2\, 0.459 ,$$

and

$$r(\hat{\boldsymbol{\beta}}_c) = \sigma^2\, 0.404 + c^2 \cdot 0.048 .$$

Therefore the difference between the two risks is

$$r(\hat{\boldsymbol{\beta}}) - r(\hat{\boldsymbol{\beta}}_c) = \sigma^2(0.055 - c^2/\sigma^2 \cdot 0.048) .$$

For $c \leqq 2\sigma^2$ we have

$$r(\hat{\boldsymbol{\beta}}) - r(\hat{\boldsymbol{\beta}}_c) \geqq \sigma^2(0.055 - c \cdot 0.096) .$$

For example when $c = 0.1$

$$r(\hat{\boldsymbol{\beta}}) - r(\hat{\boldsymbol{\beta}}_{0.1}) \geqq \sigma^2 \cdot 0.0454 .$$

This expression gives the amount by which the improved estimator has reduced the risk below that provided by the OLS estimator.

10.5 Two-stage minimax linear estimation

10.5.1 Introduction. This section considers a generalization of the minimax principle which combines two types of constraints.

First, consider the stochastic linear restriction already given in (6.1.1), namely

$$r = R\beta + d\,, \quad d \sim (0, V)\,. \tag{10.5.1}$$

As outlined in Section 6.2, this restriction may be combined with the linear model given by (10.2.1) by writing (10.5.1) and (10.2.1) as

$$\begin{pmatrix} y \\ r \end{pmatrix} = \begin{pmatrix} X \\ R \end{pmatrix} \beta + \begin{pmatrix} u \\ d \end{pmatrix}.$$

Alternatively (see Section 6.2) it may be written as

$$\tilde{y} = \tilde{X}\beta + \tilde{u}\,, \quad \tilde{u} \sim (0, \sigma^2\Phi)\,, \tag{10.5.2}$$

where

$$\Phi = \begin{pmatrix} W & 0 \\ 0 & \sigma^{-2}V \end{pmatrix}$$

and $rank\,\tilde{X} = p$. Now from Theorem 6.1 we know that the MIMSEE of β in (10.5.2) is

$$\hat{\beta}_4 = (\sigma^{-2}S + R'V^{-1}R)^{-1}\,(\sigma^{-2}X'W^{-1}y + R'V^{-1}r) = \hat{\beta}_4(\sigma^{-2})\,, \quad \text{say.} \tag{10.5.3}$$

This has

$$bias\;\hat{\beta}_4(\sigma^{-2}) = 0\,,$$

$$variance\;\hat{\beta}_4(\sigma^{-2}) = V_4 = (\sigma^{-2}S + R'V^{-1}R)^{-1} = M_{\sigma^{-2}}\,, \quad \text{say,} \tag{10.5.4}$$

and risk

$$R[\hat{\beta}_4(\sigma^{-2})] = tr\,AM_{\sigma^{-2}}\,. \tag{10.5.5}$$

Comparing $\hat{\beta}_4(\sigma^{-2})$ with the GLSE b we know from (6.2.13) that $V(b) - V_4$ is nonnegative definite.

To make $\hat{\beta}_4(\sigma^{-2})$ practicable we may replace σ^{-2} by a nonstochastic value $c \geqq 0$ giving

$$\hat{\beta}_4(c) = M_c(cX'W^{-1}y + R'V^{-1}r)\,, \tag{10.5.6}$$

where

$$M_c = (cS + R'V^{-1}R)^{-1}\,.$$

Then

$$V(\hat{\beta}_4(c)) = M_c(c^2\sigma^2S + R'V^{-1}R)\,M_c$$

and

$$R[\hat{\beta}_4(c)] = tr\,AV(\hat{\beta}(c))\,. \tag{10.5.7}$$

That is, the estimator $\hat{\beta}_4(\sigma_1^{-2})$ can be shown to be better than the GLSE for fixed σ^2 if we have prior information of the type

$$0 < \sigma_1^2 < \sigma^2 < \sigma_2^2 < \infty$$

and if certain algebraical conditions are fulfilled (see TOUTENBURG 1976).

The same principle may also be used in the minimax procedure, which will now be discussed.

10.5.2 Minimax linear estimation in the restricted model. We assume constraints of the type

$$\mathfrak{B} = \{\boldsymbol{\beta}: \boldsymbol{\beta}'\boldsymbol{T}\boldsymbol{\beta} \leqq k\} \tag{10.5.8}$$

as given in (10.2.6).

The restricted model (10.5.2) fulfills the conditions of Theorem 10.3. Therefore the MMLE of $\boldsymbol{\beta}$ may be deduced from Theorem 10.3 immediately. If we define the matrix $\boldsymbol{D}_{\sigma^{-2}}$ as

$$\boldsymbol{D}_{\sigma^{-2}} = (k^{-1}\boldsymbol{T} + \boldsymbol{M}_{\sigma^{-2}}^{-1})^{-1} ,$$

where $\boldsymbol{M}_{\sigma^{-2}}$ is given by (10.5.4), then we get the MMLE of $\boldsymbol{\beta}$ as

$$\hat{\boldsymbol{\beta}}_1(\sigma^{-2}) = \boldsymbol{D}_{\sigma^{-2}}(\sigma^{-2}\boldsymbol{X}'\boldsymbol{W}^{-1}\boldsymbol{y} + \boldsymbol{R}'\boldsymbol{V}^{-1}\boldsymbol{r}) \tag{10.5.9}$$

with

$$\textit{bias}\,[\hat{\boldsymbol{\beta}}_1(\sigma^{-2})] = -k^{-1}\boldsymbol{D}_{\sigma^{-2}}\boldsymbol{T}\boldsymbol{\beta}$$

and

$$\textit{variance}\,[\hat{\boldsymbol{\beta}}_1(\sigma^{-2})] = \boldsymbol{V}[\hat{\boldsymbol{\beta}}_1(\sigma^{-2})] = \boldsymbol{D}_{\sigma^{-2}}\boldsymbol{M}_{\sigma^{-2}}^{-1}\boldsymbol{D}_{\sigma^{-2}} .$$

Therefore,

$$\begin{aligned} r[\hat{\boldsymbol{\beta}}_1(\sigma^{-2})] &= \sup_{\boldsymbol{\beta}'\boldsymbol{T}\boldsymbol{\beta} \leqq k} R[\hat{\boldsymbol{\beta}}_1(\sigma^{-2}), \boldsymbol{\beta}, \boldsymbol{a}\boldsymbol{a}'] \\ &= \boldsymbol{a}'\boldsymbol{D}_{\sigma^{-2}}\boldsymbol{a} \\ &\leqq \boldsymbol{a}'\boldsymbol{M}_{\sigma^{-2}}\boldsymbol{a} = r[\hat{\boldsymbol{\beta}}_4(\sigma^{-2})] \\ &\leqq \sigma^2\boldsymbol{a}'\boldsymbol{S}^{-1}\boldsymbol{a} = r(\boldsymbol{b}) . \end{aligned} \tag{10.5.10}$$

In other words, whenever (10.5.8) holds, the MMLE is better than the GLSE.

10.5.3 Prior information on σ^2. The MMLE contains the unknown σ and therefore it is not practicable. The first way to overcome this is to replace σ^{-2} by a nonnegative constant $c \geqq 0$. Writing $\boldsymbol{M}_c = (c\boldsymbol{S} + \boldsymbol{R}'\boldsymbol{V}^{-1}\boldsymbol{R})^{-1}$ and $\boldsymbol{D}_c = (k^{-1}\boldsymbol{T} + \boldsymbol{M}_c^{-1})^{-1}$ this gives the modified or two-stage MML estimator

$$\hat{\boldsymbol{\beta}}_1(c) = \boldsymbol{D}_c(c\boldsymbol{X}'\boldsymbol{W}^{-1}\boldsymbol{y} + \boldsymbol{R}'\boldsymbol{V}^{-1}\boldsymbol{r}) \tag{10.5.11}$$

which has

$$bias\,[\hat{\boldsymbol{\beta}}_1(c)] = -k^{-1}\boldsymbol{D}_c\boldsymbol{T}\boldsymbol{\beta}\,,$$

$$\sup_{\boldsymbol{\beta}'\boldsymbol{T}\boldsymbol{\beta}\le k} \boldsymbol{a}'\left(bias\,\hat{\boldsymbol{\beta}}_1(c)\right)\left(bias\,\hat{\boldsymbol{\beta}}_1(c)\right)'\boldsymbol{a} = k^{-1}\boldsymbol{a}'\boldsymbol{D}_c\boldsymbol{T}\boldsymbol{D}_c\boldsymbol{a}$$

and

$$variance\,[\hat{\boldsymbol{\beta}}_1(c)] = \boldsymbol{V}\left(\hat{\boldsymbol{\beta}}_1(c)\right) = \boldsymbol{D}_c[\boldsymbol{M}_c^{-1} + (c^2\sigma^2 - c)\,\boldsymbol{S}]\,\boldsymbol{D}_c\,.$$

Therefore

$$r[\hat{\boldsymbol{\beta}}_1(c)] = \boldsymbol{a}'\boldsymbol{D}_c[c^2\sigma^2\boldsymbol{S} + k^{-1}\boldsymbol{T} + \boldsymbol{R}'\boldsymbol{V}^{-1}\boldsymbol{R}]\,\boldsymbol{D}_c\boldsymbol{a}\,. \tag{10.5.12}$$

From this we deduce that

$$\frac{\partial}{\partial c}\, r[\hat{\boldsymbol{\beta}}_1(c)] = 2(c\sigma^2 - 1)\,\boldsymbol{a}'\boldsymbol{D}_c[\boldsymbol{S} - c\boldsymbol{S}\boldsymbol{D}_c\boldsymbol{S}]\,\boldsymbol{D}_c\boldsymbol{a}\,.$$

Since $\boldsymbol{S} - c\boldsymbol{S}\boldsymbol{D}_c\boldsymbol{S}$ is nonnegative definite it follows that

$$\boldsymbol{a}'\boldsymbol{D}_c[\boldsymbol{S} - c\boldsymbol{S}\boldsymbol{D}_c\boldsymbol{S}]\,\boldsymbol{D}_c\boldsymbol{a} \geqq 0\,.$$

Therefore, using (10.5.9), $r[\hat{\boldsymbol{\beta}}_1(c)]$ is minimized for $c = \sigma^{-2}$ and $r[\hat{\boldsymbol{\beta}}_1(c)]$ is monotonically decreasing in c^{-1} if $c^{-1} < \sigma^2$.

Now suppose that we have a number σ_1^2 such that

$$0 < \sigma_1^2 < \sigma^2\,. \tag{10.5.13}$$

Then the choice $c = \sigma_1^{-2}$ in $\boldsymbol{\beta}_1(c)$ is optimal in the sense that

$$\min_{c^{-1}\geqq c_1^2} \{r[\boldsymbol{\beta}_1(c)] - r[\hat{\boldsymbol{\beta}}_1(\sigma^{-2})]\} = r[\hat{\boldsymbol{\beta}}_1(\sigma_1^{-2})] - r[\hat{\boldsymbol{\beta}}_1(\sigma^{-2})] \geqq 0\,.$$

The question now arises as to what happens if we combine the two equations $\boldsymbol{\beta}'\boldsymbol{T}\boldsymbol{\beta} \leqq k$ and $\sigma_1^2 < \sigma^2$. This gives the weaker restriction

$$\boldsymbol{\beta}'\boldsymbol{T}\boldsymbol{\beta} \leqq k^{-1}\sigma_1^{-2}\sigma^2\,. \tag{10.5.14}$$

Suppose that the covariance matrix of $\boldsymbol{d}$ is $\boldsymbol{V} = \sigma^2\tilde{\boldsymbol{V}}$. Then the 2-stage MML-estimator (10.5.11) with $c = \sigma_1^{-2}$ is

$$\hat{\boldsymbol{\beta}}_1(\sigma_1^{-2}) = (k^{-1}\sigma_1^{-2}\boldsymbol{T} + \boldsymbol{S} + \boldsymbol{R}'\tilde{\boldsymbol{V}}^{-1}\boldsymbol{R})^{-1}\,(\sigma_1^{-2}\boldsymbol{X}'\boldsymbol{W}^{-1}\boldsymbol{y} + \boldsymbol{R}'\tilde{\boldsymbol{V}}^{-1}\boldsymbol{r})\,.$$

This coincides with the MMLE under the restriction (10.5.14). In the general case where $\boldsymbol{V} \neq \sigma^2\tilde{\boldsymbol{V}}$ there is no gain in this method because the MMLE under (10.5.14) contains the unknown σ^{-2} again. That is

$$\boldsymbol{\beta}^* = (k^{-1}\sigma_1^{-2}\boldsymbol{T} + \boldsymbol{S} + \sigma^{-2}\boldsymbol{R}'\boldsymbol{V}^{-1}\boldsymbol{R})^{-1}\,(\sigma_1^{-2}\boldsymbol{X}'\boldsymbol{W}^{-1}\boldsymbol{y} + \sigma^{-2}\boldsymbol{R}'\boldsymbol{V}^{-1}\boldsymbol{r})\,.$$

Theorem 10.7 Let the covariance matrix of $\boldsymbol{d}$ be $\boldsymbol{V} = \sigma^2\tilde{\boldsymbol{V}}$, where $\tilde{\boldsymbol{V}}$ is known, and let the restrictions $\boldsymbol{\beta}'\boldsymbol{T}\boldsymbol{\beta} \leqq k$ and $\sigma_1^2 < \sigma^2$ be given. Then the two-stage MMLE, whose first stage is the MMLE under $\boldsymbol{\beta}'\boldsymbol{T}\boldsymbol{\beta} \leqq k$ and whose second stage is to replace the unknown σ^2 by σ_1^2, is identical with the MMLE under the weaker combined restriction $\boldsymbol{\beta}'\boldsymbol{T}\boldsymbol{\beta} \leqq k\sigma_1^{-2}\sigma^2$.

10.5.4 Comparison with the unrestricted GLSE. The modified MMLE estimator defined by (10.5.11) is of practical use if it gives estimators which are better than standard procedures such as the GLSE $\boldsymbol{b} = \boldsymbol{S}^{-1}\boldsymbol{X}'\boldsymbol{W}^{-1}\boldsymbol{y}$. This has $\boldsymbol{V}(\boldsymbol{b}) = \sigma^2\boldsymbol{S}^{-1}$ and $r(\boldsymbol{b}) = \boldsymbol{a}'\boldsymbol{V}(\boldsymbol{b})\,\boldsymbol{a} \geqq r[\hat{\boldsymbol{\beta}}_1(\sigma^{-2})]$.

Now we may seek a value c in such a way that

$$r(\boldsymbol{b}) - r[\hat{\boldsymbol{\beta}}_1(c)] \geqq 0\,. \tag{10.5.15}$$

Equivalently, following (10.5.12) and putting $\boldsymbol{L} = \boldsymbol{R}'\boldsymbol{V}^{-1}\boldsymbol{R} + k^{-1}\boldsymbol{T}$, we may seek to make the following quadratic forms nonnegative:

$$\boldsymbol{a}'\{\sigma^2\boldsymbol{S}^{-1} - \boldsymbol{D}_c(c^2\sigma^2\boldsymbol{S} + \boldsymbol{L})\,\boldsymbol{D}_c\}\,\boldsymbol{a}\,,$$

$$\boldsymbol{a}'\boldsymbol{D}_c\{\sigma^2\boldsymbol{D}_c^{-1}\boldsymbol{S}^{-1}\boldsymbol{D}_c^{-1} - c^2\sigma^2\boldsymbol{S} - \boldsymbol{L}\}\,\boldsymbol{D}_c\boldsymbol{a}$$

and

$$\boldsymbol{a}'\boldsymbol{D}_c\boldsymbol{L}[\sigma^2\boldsymbol{S}^{-1} + (2c\sigma^2 - 1)\,\boldsymbol{L}^{-1} + k^{-1}\boldsymbol{L}^{-1}\boldsymbol{T}\boldsymbol{L}^{-1}]\,\boldsymbol{L}\boldsymbol{D}_c\boldsymbol{a}\,.$$

Now, let $\boldsymbol{B}(c;\sigma^2)$ be the matrix in square brackets. Then we may define the class of admissible 2-stage MMLE as the set of estimators $\hat{\boldsymbol{\beta}}_1(c)$ such that the third quadratic form given above is nonnegative, namely

$$\mathfrak{K}_c = \{\hat{\boldsymbol{\beta}}_1(c);\ c \text{ such that } \boldsymbol{a}'\boldsymbol{D}_c\boldsymbol{L}\boldsymbol{B}(c;\sigma^2)\,\boldsymbol{D}_c\boldsymbol{L}\boldsymbol{a} \geqq 0\}\,.$$

We examine this problem in two stages, according to whether or not $\boldsymbol{B}$ is negative definite.

Case 1 $\boldsymbol{B}(0;\sigma^2) = \sigma^2\boldsymbol{S}^{-1} - \boldsymbol{L}^{-1} + k^{-1}\boldsymbol{L}^{-1}\boldsymbol{T}\boldsymbol{L}^{-1}$ negative definite.

Since $\boldsymbol{B}\left(\dfrac{\sigma^{-2}}{2};\sigma^2\right) = \sigma^2\boldsymbol{S}^{-1} + k^{-1}\boldsymbol{L}^{-1}\boldsymbol{T}\boldsymbol{L}^{-1}$ is positive definite, $\boldsymbol{B}(0;\sigma^2)$ is negative definite and $\boldsymbol{a}'\boldsymbol{D}_c\boldsymbol{L}\boldsymbol{B}(c;\sigma^2)\,\boldsymbol{D}_c\boldsymbol{L}\boldsymbol{a}$ is continuous in c for any $\boldsymbol{a}$ there must exist a critical value c_0, depending on $\boldsymbol{a}$, such that

$$\boldsymbol{a}'\boldsymbol{D}_c\boldsymbol{L}\boldsymbol{B}(c_0;\sigma^2)\,\boldsymbol{D}_c\boldsymbol{L}\boldsymbol{a} = 0\,,$$

$$0 < c_0 < \sigma^{-2}/2$$

and

$$\boldsymbol{a}'\boldsymbol{D}_c\boldsymbol{L}\boldsymbol{B}(c;\sigma^2)\,\boldsymbol{D}_c\boldsymbol{L}\boldsymbol{a} \geqq 0 \qquad \text{for } c > c_0\,.$$

The critical value c_0 is a function of σ^2 as well as $\boldsymbol{a}$ and therefore is unknown. But we can use prior information of the type $0 < \sigma_1^2 < \sigma^2 < \sigma_2^2 < \infty$. It is clear that if $\boldsymbol{B}(0;\sigma_2^2)$ is negative definite then $\boldsymbol{B}(0;\sigma^2)$ is also negative definite. This leads us to the following theorem.

Theorem 10.8 Suppose that we have prior information of the type $\sigma_1^2 < \sigma^2 < \sigma_2^2$ and let $\boldsymbol{B}(0;\sigma_2^2)$ be negative definite. If we confine ourselves to the set $\tilde{\mathfrak{K}}_c$ of practicable estimators which are contained in the set $\mathfrak{K}_c$ of admissible 2-stage MMLE, namely $\tilde{\mathfrak{K}}_c = \{\hat{\boldsymbol{\beta}}_1(c);\ c \geqq \sigma_1^{-2}\}$, then the estimator $\boldsymbol{\beta}_1(\sigma_1^{-2})$ is optimal in the sense that

$$\max_{c \geqq \sigma_1^{-2}} \{r(\boldsymbol{b}) - r[\hat{\boldsymbol{\beta}}_1(c)]\} = r(\boldsymbol{b}) - r[\hat{\boldsymbol{\beta}}_1(\sigma_1^{-2})] \geqq 0\,.$$

Case 2 $\boldsymbol{B}(0; \sigma^2)$ nonnegative definite (n.n.d.).

Then $\boldsymbol{B}(c; \sigma^2)$ is at least nonnegative definite for all $c > 0$. Since $\boldsymbol{B}(0; \sigma_1^2)$ n.n.d. implies $\boldsymbol{B}(0; \sigma^2)$ n.n.d., we have proved the following.

Theorem 10.9 If we have prior information of type $\sigma_1^2 < \sigma^2 < \sigma_2^2$ and if $\boldsymbol{B}(0; \sigma_1^2)$ is n.n.d. then the set of admissible 2-stage MMLEs is

$$\mathfrak{K}_c = \{\hat{\boldsymbol{\beta}}_1(c); c \geqq 0\} .$$

In practice we choose a value $c \in [\sigma_2^{-2}, \sigma_1^{-2}]$, e.g. a prior estimator $(\hat{\sigma}^2)^{-1}$ if it is contained in the interval $[\sigma_2^{-2}, \sigma_1^{-2}]$. TOUTENBURG (1975) derived similar conditions for $\hat{\boldsymbol{\beta}}_4(c)$ to have a smaller covariance matrix than b. The question arises which of the two 2-stage estimators $\hat{\boldsymbol{\beta}}_1(c)$ or $\hat{\boldsymbol{\beta}}_4(c)$ is better. The answer seems to depend on the model, on the prior information and mainly on c and σ^2 (see HOERL-KENNARD 1970a, for a similar problem in ridge regression). The following relationship holds:

$$r[\hat{\boldsymbol{\beta}}_4(c)] - r[\hat{\boldsymbol{\beta}}_1(c)] = \boldsymbol{a}'(\boldsymbol{M}_c - \boldsymbol{D}_c)\,\boldsymbol{a} + c(c\sigma^2 - 1)\,\boldsymbol{a}'[\boldsymbol{M}_c\boldsymbol{S}\boldsymbol{M}_c - \boldsymbol{D}_c\boldsymbol{S}\boldsymbol{D}_c]\,\boldsymbol{a} \ .$$

Of the three terms in this expression, the first is nonnegative, the factor $(c\sigma^2 - 1)$ may be positive or negative, and the matrix of the second quadratic form may be positive definite (e.g. if $\boldsymbol{T} = \boldsymbol{S}$) but in general will be indefinite. For $c = \sigma^{-2}$ we are led back to (10.5.10), but no further general results seem to be known.

10.6 Summary

This chapter has investigated recent Soviet work on minimax linear estimation, and related it to the technique of ridge regression. The minimax linear estimators were compared in Section 10.2.2 with the R-optimal estimators derived in Chapter 4. Problems associated with the fact that σ^2 is unknown were considered in Section 10.3, and a numerical example given in Section 10.5 considered two-stage minimax linear estimation, and showed (in Theorem 10.7) that by a suitable re-definition of constraints this can lead back to the one-stage procedure.

11

Conclusion

After all this, where do we go ?

Possibly in the theoretical direction — several unsolved problems have been pointed at, especially in the areas of estimating optimal scaling factors, and on minimax linear estimation.

Alternatively, perhaps work should now concentrate on deriving confidence regions for the '*improved*' methods advanced here. The prediction interval approach described in Chapter 8 is one way towards this. Another approach would examine the effect of using standard estimates of variance to derive confidence intervals. A third alternative would be to examine the variances of the improved estimators themselves.

A completely different line of investigation would concentrate on the practical, computational and interpretive problems associated with the methods advanced here. How are the statistics best calculated ? Can they be given some intuitive interpretation akin to least squares ? Can we really persuade people to use them ? Equally, we may well ask, *should* we persuade people to use them, or should we really advocate a continued but sceptical use of conventional methods ?

Finally, perhaps more thought should be given to the political misuses of prediction methods mentioned in Section 3.9. Do '*respectable*' statistical techniques have to be as complicated as this ? Must we technicalize the debate and thereby take it away from the democratic political arena ? It should surely be possible to find methods which are (a) clear, (b) easy to use, (c) easy to understand, and also (last *and* least) (d) not too far away from the normal canons of statistical respectability.

Appendix A

Matrix Algebra

A.1 Basic ideas

For a fuller discussion of the relevant matrix theory the reader is referred to GRAYBILL (1961), or the early chapters of JOHNSTON (1972). The present Appendix merely gives a brief summary based closely on TOUTENBURG (1975, pp. 160—170). Proofs are omitted where obvious. Theorems A 24—A 27 and A 37 on matrix differentiation are developed further in Appendix B. In what follows it is assumed that all the necessary sums, products, inverses etc. exist.

Theorem A 1 If $\boldsymbol{A}$ is an $(m \times n)$ matrix and $\boldsymbol{x}$ an $(n \times 1)$ vector, then

$$\boldsymbol{A}'\boldsymbol{A}\boldsymbol{x} = \boldsymbol{0} \quad \text{iff} \quad \boldsymbol{A}\boldsymbol{x} = \boldsymbol{0}\,.$$

Proof Clearly $\boldsymbol{A}\boldsymbol{x} = \boldsymbol{0}$ implies that $\boldsymbol{A}'\boldsymbol{A}\boldsymbol{x} = \boldsymbol{0}$. To show the converse note that

$$\boldsymbol{A}'\boldsymbol{A}\boldsymbol{x} = \boldsymbol{0} \Rightarrow \boldsymbol{y}'\boldsymbol{y} = \boldsymbol{0}\,, \qquad \text{where } \boldsymbol{y} = \boldsymbol{A}\boldsymbol{x}$$

and this implies that $\boldsymbol{y} = \boldsymbol{0}$.

Theorem A 2 If $\boldsymbol{V}$ is a vector subspace of E^T and $\boldsymbol{P}$ is the orthogonal projection of E^T into $\boldsymbol{V}$, then the codomain of $\boldsymbol{P}$ is $\boldsymbol{V}$.

A.2 Positive definiteness, etc.

Definition A 3 A matrix $\boldsymbol{A}$ is positive definite when it is symmetric and satisfies $\boldsymbol{x}'\boldsymbol{A}\boldsymbol{x} > 0$ for all $\boldsymbol{x} \neq \boldsymbol{0}$.

Theorem A 4 If $\boldsymbol{A}$ is positive definite then $\boldsymbol{A}$ has full rank, all its eigenvalues are positive, $tr\,\boldsymbol{A} > 0$, and $|\boldsymbol{A}| > 0$.

Theorem A 5 If $\boldsymbol{A}$ is positive definite and $\boldsymbol{P}$ is any commensurable matrix of full rank, then $\boldsymbol{P}'\boldsymbol{A}\boldsymbol{P}$ is positive definite. In particular (taking $\boldsymbol{A} = \boldsymbol{I}$), the matrix $\boldsymbol{P}'\boldsymbol{P}$ is positive definite for all matrices $\boldsymbol{P}$ of full rank.

Theorem A 6 If $\boldsymbol{A}$ is positive definite then $\boldsymbol{A}^{-1}$ is positive definite.

Proof Put $\boldsymbol{P} = \boldsymbol{A}^{-1}$ in Theorem A 5.

Definition A 7 A matrix $\boldsymbol{A}$ is nonnegative definite when it is symmetric and satisfies

$$\boldsymbol{x}'\boldsymbol{A}\boldsymbol{x} \geqq 0 \quad \text{for all } \boldsymbol{x} .$$

Theorem A 8 If $\boldsymbol{A}$ is positive definite then $\boldsymbol{P}'\boldsymbol{A}\boldsymbol{P}$ is nonnegative definite for all $\boldsymbol{P}$. If $\boldsymbol{P}$ is square and of full rank then $\boldsymbol{P}'\boldsymbol{A}\boldsymbol{P}$ is positive definite.

Proof Since $\boldsymbol{x}'\boldsymbol{P}'\boldsymbol{A}\boldsymbol{P}\boldsymbol{x} = \boldsymbol{y}'\boldsymbol{A}\boldsymbol{y}$ where $\boldsymbol{y} = \boldsymbol{P}\boldsymbol{x}$, then $\boldsymbol{x}'\boldsymbol{P}'\boldsymbol{A}\boldsymbol{P}\boldsymbol{x} \geqq 0$, for all $\boldsymbol{x}$. If $\boldsymbol{P}$ is square and of full rank then $\boldsymbol{y} = \boldsymbol{0}$ only when $\boldsymbol{x} = \boldsymbol{0}$, so $\boldsymbol{x}'\boldsymbol{P}'\boldsymbol{A}\boldsymbol{P}\boldsymbol{x} > 0$ for all nonzero $\boldsymbol{x}$.

Theorem A 9 For any matrix $\boldsymbol{P}$, $\boldsymbol{P}'\boldsymbol{P}$ is nonnegative definite.

Proof Put $\boldsymbol{A} = \boldsymbol{I}$ in Theorem A 8.

Theorem A 10 If $\boldsymbol{A}$ is positive definite and $\boldsymbol{B}$ is nonnegative definite, then $(\boldsymbol{A} + \boldsymbol{B})$ is positive definite.

Proof $\boldsymbol{x}'(\boldsymbol{A} + \boldsymbol{B})\,\boldsymbol{x} = \boldsymbol{x}'\boldsymbol{A}\boldsymbol{x} + \boldsymbol{x}'\boldsymbol{B}\boldsymbol{x} > 0 \quad \text{for } \boldsymbol{x} \neq \boldsymbol{0}$.

Theorem A 11 Using the notation of Theorem A 10,

(a) $|\boldsymbol{A}| \leqq |\boldsymbol{A} + \boldsymbol{B}|$,
(b) $\boldsymbol{A}^{-1} - (\boldsymbol{A} + \boldsymbol{B})^{-1}$ is nonnegative definite.

Proof See GOLDBERGER (1964, p. 38).

Definition A 12 The trace of a square matrix $\boldsymbol{A}$, $tr\,\boldsymbol{A}$, is the sum of the diagonal elements i.e.

$$tr\,\boldsymbol{A} = \sum_i a_{ii} .$$

Theorem A 13 If $\boldsymbol{A}$ and $\boldsymbol{B}$ are $(n \times n)$ matrices and c is a scalar, then

(a) $tr\,(\boldsymbol{A} + \boldsymbol{B}) = tr\,\boldsymbol{A} + tr\,\boldsymbol{B}$;
(b) $tr\,\boldsymbol{A}' = tr\,\boldsymbol{A}$;
(c) $tr\,c\boldsymbol{A} = c\,tr\,\boldsymbol{A}$;
(d) $tr\,\boldsymbol{A}\boldsymbol{B} = tr\,\boldsymbol{B}\boldsymbol{A}$.

Proof Follows directly from the definition of trace. (Part (d) is also true if $\boldsymbol{A}$ is $n \times m$ while $\boldsymbol{B}$ is $m \times n$.)

Theorem A 14 For any two $(n \times 1)$ vectors $\boldsymbol{a}$ and $\boldsymbol{b}$

$$\boldsymbol{a}'\boldsymbol{b} = \sum a_i b_i = tr\,\boldsymbol{a}\boldsymbol{b}' .$$

A.3 Idempotent matrices

Definition A 15 A matrix $\boldsymbol{A}$ is idempotent when it is symmetric and satisfies

$$\boldsymbol{A}^2 = \boldsymbol{A}\boldsymbol{A} = \boldsymbol{A} .$$

Theorem A 16 The eigenvalues of an idempotent matrix are all 1 or 0.

Theorem A 17 If $\boldsymbol{A}$ is idempotent and of full rank then $\boldsymbol{A} = \boldsymbol{I}$.

Theorem A 18 If $\boldsymbol{A}$ is idempotent and of rank r, then there exists an orthogonal matrix $\boldsymbol{P}$ such that $\boldsymbol{P}'\boldsymbol{A}\boldsymbol{P} = \boldsymbol{E}_r$, where $\boldsymbol{E}_r$ is a diagonal matrix with r ones on the main diagonal, and zeros elsewhere.

Theorem A 19 If $\boldsymbol{A}$ is idempotent and of rank r then $tr\,\boldsymbol{A} = r$.

Theorem A 20 If $\boldsymbol{A}$ and $\boldsymbol{B}$ are idempotent and $\boldsymbol{AB} = \boldsymbol{BA}$, then $\boldsymbol{AB}$ is also idempotent.

Theorem A 21 If $\boldsymbol{A}$ is idempotent and $\boldsymbol{P}$ is orthogonal, then $\boldsymbol{P}'\boldsymbol{A}\boldsymbol{P}$ is idempotent.

Theorem A 22 If $\boldsymbol{A}$ is idempotent and $\boldsymbol{B} = \boldsymbol{I} - \boldsymbol{A}$, then $\boldsymbol{B}$ is idempotent and $\boldsymbol{AB} = \boldsymbol{BA} = \boldsymbol{0}$.

Proof of Theorems A 16—A 22: see GRAYBILL (1961).

A.4 Differentiation of scalar functions of matrices

(for proofs see Appendix B, especially Figure B.2.1)

Definition A 23 If $f(\boldsymbol{X})$ is a real scalar function of an $(m \times n)$ matrix $\boldsymbol{X} = [x_{ij}]$, then the partial differential of f with respect to $\boldsymbol{X}$ is defined as the $(m \times n)$ matrix of partial differentials $\partial f/\partial x_{ij}$. That is

$$\frac{\partial f}{\partial \boldsymbol{X}} = \begin{bmatrix} \dfrac{\partial f}{\partial x_{11}} & \cdots & \dfrac{\partial f}{\partial x_{1n}} \\ \vdots & & \vdots \\ \dfrac{\partial f}{\partial x_{m1}} & \cdots & \dfrac{\partial f}{\partial x_{mn}} \end{bmatrix}.$$

Theorem A 24

(a) $\dfrac{\partial}{\partial \boldsymbol{x}} \boldsymbol{x}'\boldsymbol{A}\boldsymbol{x} = (\boldsymbol{A} + \boldsymbol{A}')\,\boldsymbol{x}$.

(b) If $\boldsymbol{A}$ is symmetric then

$$\frac{\partial}{\partial \boldsymbol{x}} \boldsymbol{x}'\boldsymbol{A}\boldsymbol{x} = 2\boldsymbol{A}\boldsymbol{x}\,.$$

Theorem A 25

$$\frac{\partial}{\partial \boldsymbol{C}} \boldsymbol{x}'\boldsymbol{C}\boldsymbol{y} = \boldsymbol{x}\boldsymbol{y}'\,.$$

Equation (B.1.2) takes particularly simple forms if y is say the trace or determinant of $\boldsymbol{Z}$. For it is easily shown (FISK 1967, pp. 147–154; ANDERSON 1958, pp. 346–349) that

$$\frac{\partial}{\partial \boldsymbol{Z}}\, tr\, \boldsymbol{Z} = \boldsymbol{I}\,, \tag{B.1.3}$$

$$\frac{\partial}{\partial \boldsymbol{Z}}\, |\boldsymbol{Z}| = |\boldsymbol{Z}|\, (\boldsymbol{Z}')^{-1}\,, \tag{B.1.4}$$

and

$$\frac{\partial}{\partial \boldsymbol{Z}}\, log\, |\boldsymbol{Z}| = (\boldsymbol{Z}')^{-1}\,. \tag{B.1.5}$$

Hence as special cases of (B.1.2) we get

$$\frac{\partial}{\partial s}\, tr\, \boldsymbol{Z} = tr\, \frac{\partial \boldsymbol{Z}}{\partial s}\,, \tag{B.1.6}$$

$$\frac{\partial}{\partial s}\, |\boldsymbol{Z}| = |\boldsymbol{Z}|\; tr\, \boldsymbol{Z}^{-1}\, \frac{\partial \boldsymbol{Z}}{\partial s}\,, \tag{B.1.7}$$

and

$$\frac{\partial}{\partial s}\, log\, |\boldsymbol{Z}| = tr\, \boldsymbol{Z}^{-1}\, \frac{\partial \boldsymbol{Z}}{\boldsymbol{Z} s}\,. \tag{B.1.8}$$

In addition matrix differentials have the usual linearity properties of differential operators, as well as the following extensions of the wellknown scalar formulae for the differential of a product:

$$\frac{\partial}{\partial \boldsymbol{X}}\, ab = \frac{\partial a}{\partial \boldsymbol{X}}\, b + a\, \frac{\partial b}{\partial \boldsymbol{X}}\,, \tag{B.1.9}$$

and

$$\frac{\partial}{\partial s}\, \boldsymbol{AB} = \frac{\partial \boldsymbol{A}}{\partial s}\, \boldsymbol{B} + \boldsymbol{A}\, \frac{\partial \boldsymbol{B}}{\partial s}\,. \tag{B.1.10}$$

B.2 The trace function

We now return for the moment to (B.1.1), which may be written as $y = y_1 + y_2$, where

$$y_1 = tr\, \boldsymbol{AX} \quad \text{and} \quad y_2 = tr\, \boldsymbol{XBX}'\,. \tag{B.2.1}$$

Let us first find the differentials $\partial y_1/\partial \boldsymbol{X}$ and $\partial y_2/\partial \boldsymbol{X}$. Clearly

$$y_1 = \sum_{\alpha,\beta} a_{\alpha\beta} x_{\beta\alpha}\,. \tag{B.2.2}$$

Therefore

$$\frac{\partial y_1}{\partial x_{ji}} = a_{ji}, \tag{B.2.3}$$

since all the terms in (B.2.2) have a zero differential, except the one where $\beta = i$ and $\alpha = j$. Now a_{ji} is the (i, j)th element of $\boldsymbol{A}'$. Hence, collecting together the results represented by (B.2.3) we deduce that

$$\frac{\partial y_1}{\partial \boldsymbol{X}} = \boldsymbol{A}' . \tag{B.2.4}$$

Having obtained this differential, we may now seek the differential of $y_2 = tr\, \boldsymbol{XBX}'$. Since this question is somewhat trickier, it will pay to be more pedantic in seeking an answer. Note that

$$y_2 = tr\, \boldsymbol{XBX}' = \sum_{\alpha, \beta, \gamma,} x_{\alpha\beta} b_{\beta\gamma} x_{\alpha\gamma} .$$

In evaluating $\partial y_2 / \partial x_{ij}$, all the terms in this summation may be ignored except when *either* $\alpha = i$ and $\beta = j$ *or* when $\alpha = i$ and $\gamma = j$. These requirements may be abbreviated in the following table

α	β	γ
i	j	—
i	—	j .

Hence the terms in y_2 which have non zero differential with respect to x_{ij} may be written

$$\sum_{\gamma} x_{ij} b_{j\gamma} x_{i\gamma} + \sum_{\beta} x_{i\beta} b_{\beta j} x_{ij} .$$

The required differential is therefore

$$\frac{\partial y_2}{\partial x_{ij}} = \sum_{\gamma} b_{j\gamma} x_{i\gamma} + \sum_{\beta} x_{i\beta} b_{\beta j} = (\boldsymbol{XB}')_{ij} + (\boldsymbol{XB})_{ij} .$$

Hence we deduce that

$$\frac{\partial y_2}{\partial \boldsymbol{X}} = \boldsymbol{XB}' + \boldsymbol{XB} .$$

Taking this equation along with (B.2.4) it is clear that the differential of y given by (B.1.1) is

$$\frac{\partial y}{\partial \boldsymbol{X}} = \frac{\partial y_1}{\partial \boldsymbol{X}} + \frac{\partial y_2}{\partial \boldsymbol{X}} = \boldsymbol{A}' + \boldsymbol{XB}' + \boldsymbol{XB} . \tag{B.2.5}$$

Turning points in the value of y are given by equating this differential to zero. The general result is given by

$$\boldsymbol{X} = -\boldsymbol{A}'(\boldsymbol{B} + \boldsymbol{B}')^{-1} .$$

When $\boldsymbol{B}$ is symmetric this equals $-\frac{1}{2}\boldsymbol{A}'\boldsymbol{B}^{-1}$, which may be compared with the value $x = -\frac{1}{2}a/b$, which minimizes the scalar analogue of (B.1.1), namely $ax + bx^2$.

An alternative way of deriving (B.2.5) is to use (B.1.2) where

$$y = tr\,\boldsymbol{Z} \quad \text{and} \quad \boldsymbol{Z} = \boldsymbol{AX} + \boldsymbol{XBX}' .$$

Now

$$\frac{\partial \boldsymbol{X}}{\partial x_{ij}} = \boldsymbol{E}_{ij} = \boldsymbol{e}_i\boldsymbol{e}_j' \quad \text{and} \quad \frac{\partial \boldsymbol{X}'}{\partial x_{ij}} = \boldsymbol{E}_{ji} = \boldsymbol{e}_j\boldsymbol{e}_i' , \tag{B.2.6}$$

where $\boldsymbol{E}_{ij}$ is the matrix with a one in the (i, j)th position and zeros elsewhere, and the vectors $\boldsymbol{e}_i$ and $\boldsymbol{e}_j$ are defined similarly. Using (B.2.6) and (B.1.10) we note that

$$\frac{\partial \boldsymbol{Z}}{\partial x_{ij}} = \boldsymbol{AE}_{ij} + \boldsymbol{E}_{ij}\boldsymbol{BX}' + \boldsymbol{XBE}_{ji} .$$

Now using (B.1.6),

$$\frac{\partial y}{\partial x_{ij}} = tr\frac{\partial \boldsymbol{Z}}{\partial x_{ij}} = tr\,(\boldsymbol{Ae}_i\boldsymbol{e}_j' + \boldsymbol{e}_i\boldsymbol{e}_j'\boldsymbol{BX}' + \boldsymbol{XBe}_j\boldsymbol{e}_i') .$$

Using the commutativity properly of the trace operator this equals

$$tr\,(\boldsymbol{e}_j'\boldsymbol{Ae}_i + \boldsymbol{e}_j'\boldsymbol{BX}'\boldsymbol{e}_i + \boldsymbol{e}_i'\boldsymbol{XBe}_j) = (\boldsymbol{A})_{ji} + (\boldsymbol{BX}')_{ji} + (\boldsymbol{XB})_{ij} .$$

This equals the (i, j)th element of $(\boldsymbol{A}' + \boldsymbol{XB}' + \boldsymbol{XB})$, thus confirming (B.2.5).

This latter method of deriving matrix differentials in fact has more general application. For instance consider

$$y = tr\,\boldsymbol{Z} , \qquad \text{where } \boldsymbol{Z} = \boldsymbol{AXBXCX}' . \tag{B.2.7}$$

Now using (B.2.6),

$$\frac{\partial \boldsymbol{Z}}{\partial x_{ij}} = \boldsymbol{AE}_{ij}\boldsymbol{BXCX}' + \boldsymbol{AXBE}_{ij}\boldsymbol{CX}' + \boldsymbol{AXBXCE}_{ji} . \tag{B.2.8}$$

Therefore from (B.1.6)

$$\frac{\partial y}{\partial x_{ij}} = tr\frac{\partial \boldsymbol{Z}}{\partial x_{ij}} .$$

But, taking the final term from (B.2.8) and using the commutativity of the trace operator we know for instance that,

$$tr\,\boldsymbol{AXBXCE}_{ji} = tr\,\boldsymbol{AXBXCe}_j\boldsymbol{e}_i' = \boldsymbol{e}_i'\boldsymbol{AXBXCe}_j = (\boldsymbol{AXBXC})_{ij} .$$

Developing the other terms in (B.2.8) similarly we see that

$$\frac{\partial y}{\partial x_{ij}} = tr\frac{\partial \boldsymbol{Z}}{\partial x_{ij}} = (\boldsymbol{BXCX}'\boldsymbol{A})_{ji} + (\boldsymbol{CX}'\boldsymbol{AXB})_{ji} + (\boldsymbol{AXBXC})_{ij} .$$

Therefore

$$\frac{\partial y}{\partial \boldsymbol{X}} = \boldsymbol{A'XC'X'B'} + \boldsymbol{B'X'A'XC'} + \boldsymbol{AXBXC} .$$

This result may be confirmed by noting the scalar valued special case, that when $y = abcx^3$ then $\partial y/\partial x = 3abcx^2$.

Further results which may be obtained in a manner similar to the above are given in Figure B.2.1. Note that Theorem A 24 is a special case of equation (b), obtained when $\boldsymbol{A}$ is symmetric and $\boldsymbol{X}$ is a vector. Theorem A 25 corresponds to (a), and Theorems A 26 and A 37 to (f).

	y	$\partial y/\partial \boldsymbol{X}$
(a)	$tr\ \boldsymbol{AX}$	$\boldsymbol{A'}$
(b)	$tr\ \boldsymbol{X'AX}$	$(\boldsymbol{A} + \boldsymbol{A'})\,\boldsymbol{X}$
(c)	$tr\ \boldsymbol{XAX}$	$\boldsymbol{X'A'} + \boldsymbol{A'X'}$
(d)	$tr\ \boldsymbol{XAX'}$	$\boldsymbol{X}(\boldsymbol{A} + \boldsymbol{A'})$
(e)	$tr\ \boldsymbol{X'AX'}$	$\boldsymbol{AX'} + \boldsymbol{X'A}$
(f)	$tr\ \boldsymbol{X'AXB}$	$\boldsymbol{AXB} + \boldsymbol{A'XB'}$

Figure B.2.1 Matrix differentials obtainable in the manner described in Section B.2.

B.3 Differentiating inverse matrices

The methods outlined in Section B.2 are insufficient to allow us to differentiate for instance

$$y = tr\ \boldsymbol{AX}^{-1} . \tag{B.3.1}$$

To do this we need the formula (B. 1.10) for differentiating a product. Clearly

$$\frac{\partial}{\partial s} \boldsymbol{T}^{-1}\boldsymbol{T} = \frac{\partial \boldsymbol{T}^{-1}}{\partial s} \boldsymbol{T} + \boldsymbol{T}^{-1} \frac{\partial \boldsymbol{T}}{\partial s} ,$$

where s is any scalar. In other words,

$$\frac{\partial \boldsymbol{T}^{-1}}{\partial s} = -\boldsymbol{T}^{-1} \frac{\partial \boldsymbol{T}}{\partial s} \boldsymbol{T}^{-1} , \tag{B.3.2}$$

because $\boldsymbol{T}^{-1}\boldsymbol{T}$ is constant, and therefore has zero differential.

This result may be extended by noting that

$$\frac{\partial}{\partial s} \boldsymbol{RT}^{-1} = \frac{\partial \boldsymbol{R}}{\partial s} \boldsymbol{T}^{-1} - \boldsymbol{RT}^{-1} \frac{\partial \boldsymbol{T}}{\partial s} \boldsymbol{T}^{-1} , \tag{B.3.3}$$

a formula which generalizes the well known scalar result

$$\frac{\partial}{\partial x}\left(\frac{u}{v}\right) = \frac{v \dfrac{\partial u}{\partial x} - u \dfrac{\partial v}{\partial x}}{v^2} .$$

As a particular case of (B.3.2) note that

$$\frac{\partial \boldsymbol{X}^{-1}}{\partial x_{ij}} = -\boldsymbol{X}^{-1}\frac{\partial \boldsymbol{X}}{\partial x_{ij}}\boldsymbol{X}^{-1} = -\boldsymbol{X}^{-1}\boldsymbol{E}_{ij}\boldsymbol{X}^{-1}\,. \tag{B.3.4}$$

Therefore, differentiating (B.3.1),

$$\frac{\partial}{\partial x_{ij}}\, tr\, \boldsymbol{A}\boldsymbol{X}^{-1} = tr\, \boldsymbol{A}\frac{\partial \boldsymbol{X}^{-1}}{\partial x_{ij}} = -\, tr\, \boldsymbol{A}\boldsymbol{X}^{-1}\boldsymbol{E}_{ij}\boldsymbol{X}^{-1} = -\,(\boldsymbol{X}^{-1}\boldsymbol{A}\boldsymbol{X}^{-1})_{ji}\,.$$

Hence we have the result that

$$\frac{\partial}{\partial \boldsymbol{X}}\, tr\, \boldsymbol{A}\boldsymbol{X}^{-1} = -\,(\boldsymbol{X}^{-1}\boldsymbol{A}\boldsymbol{X}^{-1})'\,.$$

Other results which may be derived in a similar manner are given in Figure B.3.1.

y	$\partial y/\partial \boldsymbol{X}$
$tr\, \boldsymbol{A}\boldsymbol{X}^{-1}$	$-\,(\boldsymbol{X}^{-1}\boldsymbol{A}\boldsymbol{X}^{-1})'$
$tr\, \boldsymbol{X}^{-1}\boldsymbol{A}\boldsymbol{X}^{-1}\boldsymbol{B}$	$-\,(\boldsymbol{X}^{-1}\boldsymbol{A}\boldsymbol{X}^{-1}\boldsymbol{B}\boldsymbol{X}^{-1} + \boldsymbol{X}^{-1}\boldsymbol{B}\boldsymbol{X}^{-1}\boldsymbol{A}\boldsymbol{X}^{-1})'$

Figure B.3.1 Matrix differentials involving the inverse matrix

B.4 The determinant

Suppose that we seek

$$\frac{\partial}{\partial \boldsymbol{X}}\, log\, |\boldsymbol{X}'\boldsymbol{A}\boldsymbol{X}|\,.$$

Writing $\boldsymbol{Z} = \boldsymbol{X}'\boldsymbol{A}\boldsymbol{X}$ we know that

$$\frac{\partial \boldsymbol{Z}}{\partial x_{ij}} = \boldsymbol{E}_{ji}\boldsymbol{A}\boldsymbol{X} + \boldsymbol{X}'\boldsymbol{A}\boldsymbol{E}_{ij}\,.$$

y	$\partial y/\partial \boldsymbol{X}$
$\lvert\boldsymbol{X}\rvert$	$\lvert\boldsymbol{X}\rvert\,(\boldsymbol{X}')^{-1}$
$log\,\lvert\boldsymbol{X}\rvert$	$(\boldsymbol{X}')^{-1}$
$log\,\lvert\boldsymbol{X}'\boldsymbol{A}\boldsymbol{X}\rvert$	$\boldsymbol{A}\boldsymbol{X}(\boldsymbol{X}'\boldsymbol{A}'\boldsymbol{X})^{-1} + \boldsymbol{A}'\boldsymbol{X}(\boldsymbol{X}'\boldsymbol{A}\boldsymbol{X})^{-1}$
$log\,\lvert\boldsymbol{X}'\boldsymbol{X}\rvert$	$2\boldsymbol{X}(\boldsymbol{X}'\boldsymbol{X})^{-1}$
$log\,\lvert\boldsymbol{Z}\rvert$ where $\boldsymbol{Z} = \boldsymbol{X}'\boldsymbol{A}\boldsymbol{X} + \boldsymbol{X}\boldsymbol{B} + \boldsymbol{C}\boldsymbol{X}' + \boldsymbol{D}$	$\boldsymbol{A}\boldsymbol{X}\boldsymbol{Z}^{-1} + \boldsymbol{A}'\boldsymbol{X}(\boldsymbol{Z}^{-1})' + (\boldsymbol{Z}^{-1})\,\boldsymbol{B}' + \boldsymbol{Z}^{-1}\boldsymbol{C}$

Figure. B.4.1 Matrix differentials involving the determinant function

Therefore, substituting in (B.1.8)

$$\frac{\partial}{\partial x_{ij}} \log |\boldsymbol{X}'\boldsymbol{A}\boldsymbol{X}| = tr\,(\boldsymbol{Z}')^{-1}\,[\boldsymbol{E}_{ji}\boldsymbol{A}\boldsymbol{X} + \boldsymbol{X}'\boldsymbol{A}\boldsymbol{E}_{ij}]$$

$$= [\boldsymbol{A}\boldsymbol{X}(\boldsymbol{Z}')^{-1} + \boldsymbol{A}'\boldsymbol{X}\boldsymbol{Z}^{-1}]_{ij}\,,$$

where $\boldsymbol{Z} = \boldsymbol{X}'\boldsymbol{A}\boldsymbol{X}$. In particular, when $\boldsymbol{A}$ is symmetric, this simplifies to

$$2\boldsymbol{A}\boldsymbol{X}(\boldsymbol{X}'\boldsymbol{A}\boldsymbol{X})^{-1}\,.$$

Further results involving the determinant function are given in Figure B.4.1.

Bibliography

[The numbers in square brackets at the end of each reference denote the section, or sections, in which the reference is cited.]

Aitchison, J. and I. R. Dunsmore (1968) Linear-loss interval estimation of location and scale parameters. Biometrika **55**, pp. 141—148. [8.2]

Aitchison, J. and A. R. Thatcher (1964) Two papers on the comparison of Bayesian and frequentist approaches to statistical problems of prediction (with discussion). Journal of the Royal Statistical Society B**26**(2), pp. 161—210. [1.4]

Amemiya, T. (1966) Specification analysis in the estimation of parameters of a simultaneous equation model with autoregressive residuals. Econometrica **34**, pp. 283—306. [9.5]

Anderson, T. W. (1958) An Introduction to Multivariate Statistical Analysis. Wiley. [3.8, 6.6, App. B]

Anderson, T. W. (1976) Estimation of linear functional relationships: approximate distributions and connections with simultaneous equations in econometrics (with discussion). Journal of the Royal Statistical Society B **38**, pp. 1—36. [3.1]

Anscombe, F. (1973) Graphs in statistical analysis. The American Statistician 27(1), February, pp. 17—21. [3.7]

Appa, G. and C. Smith (1973) On L_1 and Chebyshev estimation. Mathematical Programming 5(1), pp. 73—87. [3.2]

Bacon, R. W. and J. A. Hausman (1974) The relationship between ridge regression and the minimum mean square error estimator of Chipman.

Bulletin of the Oxford University Institute of Economics and Statistics **36**(2), pp. 115—124. [2.3]

Bibby, J. (1972) Minimum mean square error estimation, ridge regression, and some unanswered questions. Proceedings of the 9th European Meeting of Statisticians, Budapest, reprinted in J. Gani, K. Sarkadi and I. Vincze, eds. (1974) Progress in Statistics. Colloquia Mathematica Janos Bolyai, pp. 107—121. [1.3, 2.2, 2.4, 4.2, 7.5, 10.2]

Bibby, J. (1974) On lines and scattergrams. Mimeo. [3.1]

Bibby, J. (1977) The general linear model — a cautionary tale. Pp. 35—79 in C. O'Muircheartaigh and C. Payne, eds. Model fitting: The Analysis of Survey Data, Vol. 2. Wiley. [3.1, 3.4, 3.5, 3.7, 3.9]

Blight, B. J. N. (1971) Some general results on reduced mean square error estimation. The American Statistician **25**(3), June, pp. 24—25. [2.1, 2.3]

Bowles, S. and H. M. Levin (1968) The determinants of scholastic achievement: an appraisal of some recent evidence. Journal of Human Resources, **3**(1), pp. 3—24. [3.3]

Box, G. E. P. and G. M. Jenkins (1970) Time Series, Forecasting and Control. Holden-Day. [1.1, 1.4]

Brown, R. G. (1962) Smoothing, Forecasting and Prediction of Discrete Time Series. Prentice-Hall International. [1.1, 1.4]

Brown, P. (1974) Predicting by ridge regression. Unpublished. [1.3]

Brown, P. and C. Payne (1975) Election night forecasting (with discussion). Journal of the Royal Statistical Society A138, pp. 463—498. [1.3]

Bunke, O. (1963) Theorie der Bereichschätzung Statistischer Parameter (Interval estimation theory for statistical parameters). Teorija verojatnostej i ee primenenija 8(3), pp. 330—337. [8.5]

Chipman, J. S. (1964) On least squares with insufficient observations. Journal of the American Statistical Association **59**, pp. 1078—1111. [2.3]

Coleman, J. S. et al. (1966) Equality of Educational Opportunity: U.S. Government Printing Office. [3.3]

Covey-Crump, P. A. K. and S. D. Silvey (1970) Optimal regression designs with previous observations. Biometrika **57**, pp. 551—566. [5.1]

Cox, D. R. and E. J. Snell (1971) On test statistics calculated from residuals. Biometrika, **58**(3), pp. 589—594. [3.6]

Cramér, H. (1966) Mathematical Methods of Statistics. Princeton University Press. [9.1]

Deegan, J., Jr. (1974) Specification error in causal models. Social Science Research **3**(3), September, pp. 235—259. [3.5]

Dhrymes, Phoebus J. (1970) Econometrics: Statistical Foundations and Applications. Harper & Row. [3.1, 3.8]

Draper, N. and H. Smith (1966) Applied Regression Analysis. John Wiley & Sons, Inc. [3.5, 3.6, 3.8]

Draper, N. and H. Smith (1969) Methods for selecting variables from a given set of variables for regression analysis (with discussion). Bulletin of the International Statistical Institute **43**(1), pp. 7—15. [3.6]

Duncan, O. D., A. O. Haller and A. Portes (1968) Peer influences on aspirations: a reinterpretation. American Journal of Sociology, **74**(2), pp. 119—137. [3.2, 3.9]

Efron, B. and C. Morris (1973) Combining possibly related estimation problems (with discussion). Journal of the Royal Statistical Society B**35**(3), pp. 379—421. [1.3]

Farebrother, R. W. (1975) The minimum mean square error linear estimator and ridge regression. Technometrics **17**(1), pp. 127—128. [4.2]

Finney, D. J. (1974) Problems, data and inference. Journal of the Royal Statistical Society A**137**(1), pp. 1—23. [1.3]

Finney, D. J. (1975) Numbers and data. Biometrics **31**(2), pp. 375—386. [1.3]

Fisk, P. R. (1967) Stochastically Dependent Equations: An Introductory Text for Econometricians. Griffin. [App. B]

Geary, R. C. (1947) Testing for normality. Biometrika **34**, pp. 209—242. [3.2]

Goldberger, A. S. (1962) Best linear unbiased prediction in the generalized linear regression model. Journal of the American Statistical Association **57**, pp. 369—375. [1.5, 5.1, 5.2]

Goldberger, A. S. (1964) Econometric Theory. Wiley. [3.1, 3.4, 3.8, 4.4, 9.5, 10.4, App. A]

Goldstein, M. and A. F. M. Smith (1974) Ridge-type estimators for regression analysis. Journal of the Royal Statistical Society B**36**(2), pp. 284—291. [1.3]

Gordon, R. A. (1968) Issues in multiple regression. American Journal of Sociology, **73**(5), pp. 592—616. [3.3]

Graybill, F. A. (1961) An Introduction to Linear Statistical Models. McGraw-Hill. [App. A]

Guttman, I. (1970) Statistical Tolerance Regions. Griffin. [1.1, 8.2, 8.4, 8.6]

Harding, E. F. (1972) Contribution to the discussion on Lindley and Smith (1972) [1.3]

Harrison, P. J. (1965) Short-term sales forecasting. Applied Statistics **14**, pp. 102—139. [1.4]

Hemmerle, W. J. (1975) An explicit solution for generalized ridge regression. Technometrics **17**(3), August, pp. 309—314. [1.3, 2.4]

Hibbs, D. A. (1974) Problems of statistical estimation and causal inference in time-series regression models. Pp. 252—308 in H. L. Costner, ed., Sociological Methodology 1973—1974. Jossey-Bass Inc. [3.8]

Hoerl, A. E., and R. W. Kennard (1970a) Ridge regression: biased estimation for nonorthogonal problems. Technometrics **12**, pp. 55—67. [1.1, 1.3, 4.2, 6.2, 10.1]

Hoerl, A. E. and R. W. Kennard (1970b) Ridge regression: applications to nonorthogonal problems. Technometrics **12**, pp. 69—82. [1.1, 1.3, 4.2, 6.2, 10.1]

Hoerl, A. E. and R. W. Kennard (1975) A note on a power generalization of ridge regression. Technometrics **17**(2), p. 269. [1.3]

Holt, C. C. (1957) Forecasting seasonal and trends by exponentially weighted moving averages. Carnegie Institute of Technology. [1.4]

Hope, K. (1968) Methods of Multivariate Analysis. University of London Press. [3.3]

International Encyclopaedia of the Social Sciences: see Sills (1968)

James, W. and C. Stein (1961) Estimation with quadratic loss. Proceedings of the 4th Berkeley Symposium on Mathematical Statistics and Probability **1**, pp. 361—379. [1.3]

Johnston, J. (1972) Econometric Methods. McGraw-Hill (second edition). [3.1, 3.4, 3.8, App. A]

Kadiyala, K. R. (1970) An exact small sample property of the k-class estimators. Econometrica **38**, pp. 930—932. [9.5]

Kakwani, N. C. (1965) Note on the use of prior information in forecasting with a linear regression model. Sankhya A27(1), pp. 101—104. [6.2]

Kendall, M. G. (1957) A Course in Multivariate Analysis. Griffin. [3.2]

Kendall, M. G. (1973) Time-series. Griffin. [1.1, 1.4, 3.6]

Kendall, M. G. and A. Stuart (various) The Advanced Theory of Statistics. Vol. 1: Distribution Theory Vol. 2: Inference and Relationships Vol. 3: Design and Analysis, and Time-Series. Griffin. [2.1, 2.3, 2.4]

Kiountouzis, E. A. (1973) Linear programming techniques in regression analysis. Applied Statistics **22**(1), pp. 69—73. [3.2]

Kuks, J. (1972) Minimaksnaja ozenka koeffizientow regressii (A minimax estimator of regression coefficients). Iswestija Akademija Nauk Estonskoj SSR **21**, pp. 73—78. [10.2, 10.3]

Kuks, J. and W. Olman (1971) Minimaksnaja linejnaja ozenka koefficientow regressii (Minimax linear estimation of regression coefficients). Iswestija Akademija Nauk Estonskoj SSR **20**, pp. 480—482. [10.2]

Kuks, J. and W. Olman (1972) Minimaksnaja linejnaja ozenka koeffizientow regressii, II (Minimax linear estimation of regression coefficients, II). Iswestija Akademija Nauk Estonskoj SSR **21**, pp. 66—72. [10.2, 10.3]

Kupper, L. L. and E. F. Meydrech (1973) A new approach to mean squared error estimation of response surfaces. Biometrika **60**(3), pp. 573—579. [4.2]

Läuter, H. (1970) Optimale Vorhersage und Schätzung in regulären und singulären Regressionsmodellen (Optimal prediction and estimation in regular and singular regression models). Mathematische Operationsforschung und Statistik **1**, pp. 229—243. [5.4]

Läuter, H. (1975) A minimax linear estimator for linear parameters under restrictions in form of inequalities. Mathematische Operationsforschung und Statistik **5**, pp. 689–696.

Legendre, A. M. (1805) Nouvelles méthodes pur la détermination des orbites des comètes. Paris. [3.2]

Liebermann, G. L. and R. G. Miller (1963) Simultaneous tolerance intervals in regression. Biometrika **50**, pp. 155–168. [8.4]

Lindley, D. V. and A. F. M. Smith (1972) Bayes estimates for the linear model (with discussion). Journal of the Royal Statistical Society B**34**(1), pp. 1–41. [1.3, 2.3]

McElroy, F. W. (1967) A necessary and sufficient condition that ordinary least-squares estimators be best linear unbiased. Journal of the American Statistical Association **62**, pp. 1302–1304. [3.5, 4.4]

Mardia, K. V. and J. Bibby (1977) Statistics of Multivariate Data. Academic Press. [6.4]

Neter, J. and W. Wasserman (1974) Applied Linear Statistical Models. Richard D. Irwin Inc. [3.6]

Newbold, P. and C. W. J. Granger (1974) Experience with forecasting univariate time series and the combination of forecasts (with discussion). Journal of the Royal Statistical Society A**137**(2), pp. 131–164. [1.4, 2.3]

Open University (1977) M 341: Fundamentals of Statistical Inference. [3.5, 3.7]

Peaker, G. F. (1967) The regression analyses of the National Survey. App. 4 in Vol. 2 of Children and their Primary Schools, A Report of the Central Advisory Council for Education (England) (The Plowden Report). London: H.M.S.O. [3.3]

Peaker, G.F. (1971) The Plowden Children Four Years later. Slough: N.F.E.R. [3.3]

Perlman, M. D. (1972) Reduced mean square error estimation for several parameters. Sankhya B**34**(1), pp. 89–92. [2.1, 2.3]

Pigeon, D. A., ed. (1967) Achievement in Mathematics: a National Study of Secondary Schools. Slough: N.F.E.R. [3.3]

Pitman, E. J. G. (1937) The 'closest' estimates of statistical parameters. Proceedings of the Cambridge Philosophical Society **33**, pp. 212–222. [7.2]

Rao, C. R. (1971) Unified theory of linear estimation, Sankya A33, pp. 370–396 and Sankhya A**34**, p. 477. [1.2, 2.2, 4.2]

Rao, C. R. (1973) Linear Statistical Inference and Its Applications. Wiley (Second edition). [2.1, 2.2]

Rao, P. and R. L. Miller (1971) Applied Econometrics. Wadsworth Publishing Co. [3.1, 3.5, 3.6, 3.8]

Rice, J. R. and J. S. White (1964) Norms for smoothing and estimation. SIAM Review **6**(3), pp. 243–256. [3.2]

Schönfeld, P. (1969) Methoden der Ökonometrie (Econometric Methods). Berlin. [4.1]

Schuessler, K. (1968) Prediction. In Sills (1968). [1.4]

Sills, D. L., ed. (1968) International Encyclopaedia of the Social Sciences. Macmillan and The Free Press. [1.4]

Smith, A. F. M. and M. Goldstein (1975) Ridge regression: some comments on a paper of Conniffe and Stone. The Statistician **24**(1), pp. 61–66. [1.3]

Sprent, P. (1969) Models in Regression and Related Topics. Methuen. [3.6, 4.4, 4.5]

Stein, C. M. (1956) Inadmissibility of the usual estimator for the mean of a multivariate normal distribution. Proceedings of the 3rd Berkeley Symposium on Mathematical Statistics and Probability **1**, pp. 197–206. [1.1, 1.3]

Stein, C. M. (1962) Confidence sets for the mean of a multivariate normal distribution (with discussion). Journal of the Royal Statistical Society B**24**, pp. 265–296. [1.3]

Stigler, S. M. (1973) Laplace, Fisher, and the discovery of the concept of sufficiency. Biometrika 60(3), pp. 439—445. [illegible]

Stone, M. (1974) Cross validatory choice and assessment of statistical predictions. Journal of the Royal Statistical Society **B36**(2), pp. 111—147. [1.3]

Stuart, A. (1969) Reduced mean-square error estimation of σ^p in normal samples. The American Statistician **23**(4), pp. 27—28. [2.1, 2.3]

Subrahmanya, M. T. (1970) A note on non-negative estimators of positive parameters. Metrika **16**, pp. 106—114. [1.3]

Theil, H. (1961) Economic Forecasts and Policy. North Holland Press. Revised edition (First edition 1957). [5.1]

Theil, H. (1961) On the use of incomplete prior information in regression analyss. Journal of the American Statistical Association **58**, pp. 401—414. [6.1, 10.3]

Theil, H. (1966) Applied Economic Forecasting. North Holland Publishing Co. (2nd printing 1971) [1.1, 1.5, 5.1]

Theil, H. (1971) Principles of Econometrics. Wiley. [4.2]

Theil, H. and A. S. Goldberger (1961) On pure and mixed estimation in economics. International Economic Review **2**, pp. 65—78. [6.2]

Thompson, J. R. (1968) Some shrinkage techniques for estimating the mean. Journal — the American Statistical Association **63**, pp. 113—122. [2.4]

Till, R. (1973) The use of linear regression in geomorphology. Area 5(4), pp. 303—308. [3.2]

Toro-Vizcarrondo, C. E. and T. D. Wallace (1968) A test of the mean square error criterion for restrictions in linear regression. Journal of the American Statistical Association **63**, pp. 558—572. [5.5, 7.5]

Toutenburg, H. (1968) Vorhersage im allgemeinen linearen Regressionsmodell mit Zusatzinformation über die Koeffizienten (Prediction in the general linear regression model with auxiliary information on the coefficients). Operationsforschung und Mathematische Statistik, Sonderband I, pp. 107—120. [4.2, 5.3]

Toutenburg, H. (1970 a) Vorhersage im allgemeinen linearen Regressionsmodell mit stochastischen Regressoren (Prediction in the general linear regression model with stochastic regressors). Operationsforschung und Mathematische Statistik **2**, pp. 105—116. [9.1]

Toutenburg, H. (1970 b) Probleme linearer Vorhersagen im allgemeinen linearen Regressionsmodell (Problems of linear predictors in the general linear model). Biometrische Zeitschrift **12** (4), pp. 242—252. [5.4]

Toutenburg, H. (1970 c) Über die Wahl zwischen erwartungstreuen oder nicht erwartungstreuen Vorhersagen (On choosing between biased and unbiased predictors). Operationsforschung und Mathematische Statistik, Sonderband II, pp. 107—118. [9.2]

Toutenburg, H. (1971) Probleme der Intervallvorhersage von normalverteilten Variablen (Problems of interval prediction with normally distributed variables). Biometrische Zeitschrift **13** (4), pp. 261—273. [8.5]

Toutenburg H. (1973) Lineare Restriktionen und Modellwahl im allgemeinen linearen Regressionsmodell (Linear restrictions and model choice in the general linear regression model). Biometrische Zeitschrift **15** (5), pp. 325—342. [7.2]

Toutenburg, H. (1975) Vorhersage in Linearen Modellen (Prediction in Linear Models). Berlin, Akademie-Verlag. [Passim]

Tukey, J. W. (1975) Instead of Gauss-Markov least squares, what? Pp. 351—372 in R. P. Gupta, ed., Applied Statistics, Proceedings of the Conference at Dalhousie University, Halifax, May 2—4, 1974. North Holland Publishing Co. [3.4, 3.5]

Toutenburg, H. (1976) Minimax-linear and MSE-estimators in generalized regression. Biometrische Zeitschrift 18 (2), pp. 91—100. [10.3, 10.4]

Van de Geer, J. P. (1971) Introduction to Multivariate Analysis for the Social Sciences. W. H. Freeman and Company. [3.8]

Wallace, T. D. and C. E. Toro-Vizcarrondo (1969) Tables for the mean square error test for exact linear restrictions in regression. Journal of the American Statistical Association **64**, pp. 1649—1663. [5.5, 7.5]

Webster, J. T. (1965) On the use of a biased estimator in linear regression. Journal of the Indian Statistical Association **3**, pp. 82—90. [7.2]

Whittle, P. (1962) Contribution to discussion on Stein (1962). [2.4]

Wickens, M. R. (1969) The consistency and efficiency of generalized least squares in simultaneous equation systems with autocorrelated errors. Econometrica **37**, pp. 651—659. [9.5]

Winters, P. R. (1960) Forecasting sales by exponentially weighted moving averages. Management Science **6**, pp. 324—342. [1.4]

Witting, H. and G. Nölle (1970) Angewandte Mathematische Statistik (Applied Mathematical Statistics). B. G. Teubner Stuttgart. [8.6]

Wold, H. (1963) Forecasting by the chain principle. Pp. 471—497 in M. Rosenblatt, ed., Time Series Analysis, Wiley. [1.4]

Zacks, S. (1971) The Theory of Statistical Inference. Wiley. [2.3]

Zarnowitz, V. (1968) Prediction and forecasting, economic. In Sills (1968). [1.4]

Zellner, A. (1962) An efficient method for estimating seemingly unrelated regressions and tests for aggregation bias. Journal of the American Statistical Association **57**, pp. 348—368 [3.8]

Zellner, A. and W. Vandaele (1975) Bayes-Stein estimators for k-means, regression and simultaneous equation models. Pp. 627—653 in S. E. Fienberg and A. Zellner, eds., Studies in Bayesian Econometrics and Statistics in Honor of Leonard J. Savage. North-Holland. [1.3, 2.3]

Index

Glossary

This glossary presents a brief description of the notational conventions and abbreviations adopted throughout the book. For further details, the reader is referred to the index.

Notational conventions

$X \sim (\mu, \sigma^2)$	The random variable X has mean μ and variance σ^2.
$X \sim N(\mu, \sigma^2)$	The random variable X has the normal distribution with mean μ and variance σ^2.
$\boldsymbol{X} \sim (\boldsymbol{\mu}, \boldsymbol{\Sigma})$	The random vector $\boldsymbol{X}$ has mean vector $\boldsymbol{\mu}$ and variance-covariance matrix $\boldsymbol{\Sigma}$.
$\boldsymbol{X} \sim N_p(\boldsymbol{\mu}, \boldsymbol{\Sigma})$	The random vector $\boldsymbol{X}$ has the p-variate normal distribution with mean vector $\boldsymbol{\mu}$ and variance-covariance matrix $\boldsymbol{\Sigma}$.
$U(a, b)$	The (continuous) uniform distribution on the range (a, b).
χ^2_m	The (central) chisquared distribution with m degrees of freedom.
$\chi^2_m(\lambda)$	The noncentral chisquared distribution with m degrees of freedom, and noncentrality parameter λ.
$F_{i,j}$	The (central) F distribution with i and j degrees of freedom.
$F_{i,j}(\lambda)$	The noncentral F distribution with i and j degrees of freedom, and noncentrality parameter λ.

Abbreviations

$\boldsymbol{b}$	The GAUSS-MARKOV Generalized least squares estimator of $\boldsymbol{\beta}$, defined in (3.4.1) as $\boldsymbol{b} = (\boldsymbol{X}'\boldsymbol{\Sigma}^{-1}\boldsymbol{X})^{-1}\,\boldsymbol{X}'\boldsymbol{\Sigma}^{-1}\boldsymbol{y}$.
$\boldsymbol{b}_0$	A general linear estimator of the vector $\boldsymbol{\beta}$, defined in (1.2.3) as $\boldsymbol{b}_0 = \boldsymbol{A}\boldsymbol{y} + \boldsymbol{c}$.
$\boldsymbol{\beta}$	The regression coefficients in the model $\boldsymbol{y} = \boldsymbol{X}\boldsymbol{\beta} + \boldsymbol{u}$ defined in (1.2.1).

$\hat{\boldsymbol{\beta}}$	The ordinary least squares estimator of β, defined in (1.2.2) as $\hat{\boldsymbol{\beta}} = (\boldsymbol{X}'\boldsymbol{X})^{-1}\boldsymbol{X}'\boldsymbol{y}$.
BLU	Best linear unbiased.
BLUE	Best linear unbiased estimator.
c.d.f.	Cumulative distribution function, $F_X(x) = Prob(X \leqq x)$.
Covar	Covariance, covariance matrix
$\boldsymbol{e}_i$	A vector with one in the ith position and zeros elsewhere.
E	Expectation.
$\overline{E}$	Asymptotic expectation
$\boldsymbol{E}_{ij}$	A matrix with one in the (i, j)th position and zeros elsewhere, $\boldsymbol{E}_{ij} = \boldsymbol{e}_i\boldsymbol{e}_{j'}$.
$\boldsymbol{G}$	The matrix $\boldsymbol{T}^{-1} = (\boldsymbol{X}'\boldsymbol{X})^{-1}$, used in defining $\hat{\boldsymbol{\beta}} = \boldsymbol{G}\boldsymbol{X}'\boldsymbol{y}$ (see 1.2.2).
GLS	Generalized least squares. (p. 56).
GM	GAUSS-MARKOV.
$\boldsymbol{H}$	The matrix $\boldsymbol{I}_n - \frac{1}{n}\boldsymbol{1}\boldsymbol{1}'$ used in centering. $\boldsymbol{H}$ is an $(n \times n)$ matrix whose diagonal elements are $1 - \frac{1}{n}$ and whose off-diagonal elements are $-\frac{1}{n}$. $\boldsymbol{H}$ has the property that the ith element of $\boldsymbol{H}\boldsymbol{x}$ is $(x_i - \bar{x})$, where $\bar{x}$ is the mean of the x_i's (that is, $\bar{x} = \frac{1}{n}\boldsymbol{x}'\boldsymbol{1}$). (See p. 49.)
i.i.d.	Independent and identically distributed.
MIMSEE	Minimum mean square error estimator, described in Section 2.2.
MPE	Mean pth power error, defined in (2.5.3).
MSE	Mean square error, defined in (1.2.4).
MSEP	Mean square error of prediction, defined in (1.5.5).
OLS	Ordinary least squares. (See 1.2.2)
ORP	Orthogonal regression procedure — see Section 3.2.
$\boldsymbol{P}$	The idempotent matrix $\boldsymbol{X}(\boldsymbol{X}'\boldsymbol{X})^{-1}\boldsymbol{X}' = \boldsymbol{X}\boldsymbol{G}\boldsymbol{X}'$.
p.d.f.	Probability density function.
$\boldsymbol{Q}$	The idempotent matrix $\boldsymbol{I} - \boldsymbol{P}$.
RMA	Reduced major axis procedure — see Section 3.2.
RSS	Residual sum of squares.
$\boldsymbol{S}$	The matrix $\boldsymbol{X}'\boldsymbol{\Sigma}^{-1}\boldsymbol{X}$, used in defining $\boldsymbol{b} = \boldsymbol{S}^{-1}\boldsymbol{X}'\boldsymbol{\Sigma}^{-1}\boldsymbol{y}$ (see 3.4.1).
$\boldsymbol{T}$	The matrix $\boldsymbol{X}'\boldsymbol{X}$ used in (1.2.2).
TSSM, TSS	Total sum of squares about the mean (pp. 27, 49).
$\boldsymbol{V}$, *Var*	Variance, variance-covariance matrix (p. 4).